Fall Prevention through Design in Construction

T0187731

The role of designers has traditionally been to design a building so that it conforms to accepted local building codes. The safety of workers is left up to the contractor building the designs. Research shows, however, that designers can have an especially strong influence on construction safety during the conceptual, preliminary and detailed design phases.

This book establishes the new knowledge and conceptual frameworks necessary to develop a mobile computing-enabled knowledge-management system that can help reduce the high rate of construction falls. There are three main objectives of this book:

1   to create a new Prevention through Design (PtD) knowledge base to model the relationships between fall risks and design decisions;
2   to develop a PtD mobile app to assist building designers in fall PtD; and
3   to evaluate the practical implications of the PtD mobile app for the construction industry, especially for building designers and workers.

The cutting-edge technologies explored in this book have the potential to significantly reduce the rate of serious injuries that occur in the global construction industry. This is essential reading for researchers and advanced students of construction management with an interest in safety or mobile technologies.

Imriyas Kamardeen is a Senior Lecturer at UNSW, Australia, specialising in ICT in Construction Management, and also Editor of *Construction Economics and Building* (formerly known as the *Australasian Journal of Construction Economics and Building* (AJCEB)). His first book with Routledge, *OHS Electronic Management Systems for Construction*, was published in 2013.

Spon Research

publishes a stream of advanced books for built environment researchers and professionals from one of the world's leading publishers. The ISSN for the Spon Research programme is ISSN 1940-7653 and the ISSN for the Spon Research E-book programme is ISSN 1940-8005

Published:

Free-Standing Tension Structures: From Tensegrity Systems to Cable-Strut Systems
978-0-415-33595-9
W. B. Bing

Performance-Based Optimization of Structures: Theory and Applications
978-0-415-33594-2
Q. Q. Liang

Microstructure of Smectite Clays & Engineering Performance
978-0-415-36863-6
R. Pusch & R. Yong

Procurement in the Construction Industry: The Impact and Cost of Alternative Market and Supply Processes
978-0-415-39560-1
W. Hughes et al.

Communication in Construction Teams
978-0-415-36619-9
S. Emmitt & C. Gorse

Concurrent Engineering in Construction Projects
978-0-415-39488-8
C. Anumba, J. Kamara & A.-F. Cutting-Decelle

People and Culture in Construction
978-0-415-34870-6
A. Dainty, S. Green & B. Bagilhole

Very Large Floating Structures
978-0-415-41953-6
C. M. Wang, E. Watanabe & T. Utsunomiya

Tropical Urban Heat Islands: Climate, Buildings and Greenery
978-0-415-41104-2
N. H. Wong & C. Yu

Innovation in Small Construction Firms
978-0-415-39390-4
P. Barrett, M. Sexton & A. Lee

Construction Supply Chain Economics
978-0-415-40971-1
K. London

Fall Prevention through Design in Construction

The benefits of mobile computing

Imriyas Kamardeen

Routledge
Taylor & Francis Group

LONDON AND NEW YORK

British Library Cataloguing-in-Publication Data
A catalogue record for this book is available from the British Library

Library of Congress Cataloging in Publication Data
Kamardeen, Imriyas.
Fall prevention through design in construction : the benefits of mobile
computing/Imriyas Kamardeen.
pages cm
Includes bibliographical references and index.
1. Building—Safety measures. 2. Falls (Accidents)—Prevention—Data
processing. 3. Mobile computing. 4. Construction industry—Accidents—
Prevention—Data processing. 5. Architecture—Human factors—Data
processing. I. Title.
TH443.K36 2015
690′.220285535—dc23
2014048782

ISBN: 978-1-138-77915-0 (hbk)
ISBN: 978-0-367-73816-7 (pbk)

Typeset in Sabon
by Swales & Willis Ltd, Exeter, Devon, UK

Dedicated to my parents and Zakia, Mahdiyya and Imaad

Contents

Figures and exhibits

Figures

Exhibits

Tables

Preface

Falls are a leading cause of workplace fatalities and serious injuries in construction globally. They represent 30–50% of construction fatalities in developed nations. Falls are also a heavy financial burden so much so that the total annual cost of falls amounts to millions of dollars in many countries. The human sufferings and socio-psychological miseries endured by fall victims and their families are enormous and at the same time are absolutely unacceptable. Prevention and control of falls are therefore an urgent necessity in construction.

Embracing Prevention through Design (PtD) principles in design practices is recognised as an effective way to eliminate fall hazards at source and thereby curb the death toll in construction. Nonetheless, the successful adoption of PtD encounters a significant challenge, which is the knowledge and skills gap of designers. Having been educated and trained predominantly in design principles and not having worked on construction sites, many designers have limited understanding of the hazard and risk consequences of their design decisions. This situation is dangerous not only for construction operatives but also for designers themselves because workplace health and safety legislation in many countries across the globe now holds designers liable for workplace accidents. Hence, equipping designers with the necessary PtD knowledge, skills and toolkits is a pressing industry need.

In this vein, this book contributes to addressing the above need by creating a new knowledge base of safe design solutions for fall prevention, and an innovative mobile app to facilitate the application of the knowledge base in real-life design contexts. The book is based on recent research and best practices, and is capable of adding value to existing practices in the construction industry. Chapters in the book gradually deliver the theme to readers through systematic layouts of discussions. The work is easily understandable by anyone from the field of construction or outside. Moreover, the research journey that was undertaken in authoring the book is vividly explained to enable easy applications and repetitions in similar studies by other researchers.

The book will appeal equally to industry professionals as well as academic and research communities. It will significantly benefit designers in exercising their PtD obligations and thereby avoid penalties and prosecutions. Likewise, because there are no notable textbooks that academics can use for teaching accident PtD, and because this theme has become one of the threshold concepts for built environment education, this book will underpin curriculum enhancement in tertiary institutions. The book would also be of value to researchers internationally who specialise in the areas of construction health and safety management, risk management, and mobile computing in construction.

Acknowledgements

This book is the outcome of a research study conducted over a period of two years. The research might not have been successful without the support rendered by various individuals and organisations. I am morally obliged to acknowledge the different forms of support provided to me. First and foremost, I would like to thank my employer, the University of New South Wales, Australia for providing me with a research grant and other resources necessary for undertaking this research. Special thanks go to Alec Tzannes, Director of Tzannes Associates and his colleague Georgina Blix for sharing with me sample PtD documents from their projects and for participating in the mobile app scope formulation. Likewise, I would like to extend my appreciation to Nick Kodos and Wilson Fan of Ganellen Construction for offering building designs for the mobile app evaluation and for granting permission to include them in this book. Thanks to the construction industry professionals who took part in the review of the rapid mobile app prototype as well as to those who participated in the evaluation of the final mobile app. Last but not least, I would like to thank my wife Zakia Rizvi for her continued support to my career, especially during the development of this book.

Imriyas Kamardeen
University of New South Wales, Australia

Disclaimer

Whilst the author has taken reasonable care in the preparation of this book, the use of the book is at your own risk and on an 'as is' basis. The author makes no expressed or implied warranty of any kind and will not be liable to you for any errors or omissions.

The author excludes any liability to you or anyone claiming through you for any loss, liability or damage whether directly or indirectly suffered or incurred by you or a third party, arising from, or in connection with, the use of the book, including any information derived from it.

You should carefully review any recommendations or information contained in this book and obtain independent verification from a suitably qualified professional before acting on any of those recommendations. No responsibility is assumed by the author for any injury and/or damage to person or property arising from any methods, instructions or recommendations contained in this book.

Abbreviations

app	Application
BIM	Building Information Modelling
CDM	Construction Design and Management
CHAIR	Construction Hazard Assessment Implication Review
CHPtD	Construction Hazard Prevention through Design
CII	Construction Industry Institute
COF	Coefficient of Friction
DfCS	Designing for Construction Safety
DfS	Design for Safety
DSS	Decision Support System
HSE	Health and Safety Executive
HVAC	Heating, Ventilation and Air Conditioning
ICT	Information and Communication Technologies
iOS	iPhone Operating System
MEP	Mechanical, Electrical and Plumbing
NSW	New South Wales
OHS	Occupational Health and Safety
OSHA	Occupational Safety and Health Administration
PtD	Prevention through Design
SDK	Software Development Kit
WHS	Workplace Health and Safety

1 Introduction

Safety profile of construction

Accident prevention and control in construction have presented a relentless, global challenge, which has inevitably profiled the construction industry with one of the worst workplace safety statistics among other highly hazardous sectors such as mining, electrical, chemical and transportation industries (Hu et al. 2011). The US construction industry, for example, remains the major source of fatal accidents, accounting for 19% of all workplace fatalities (Bureau of Labor Statistics 2010). Similarly in the UK, construction constitutes 21.5% of all workplace fatalities (Health and Safety Executive [HSE] 2010) and records a non-fatal injury rate of 16 per 1000 workers, considerably higher than the overall industry rate of 10 per 1000 workers (Labour Force Survey 2009). The Australian construction industry's accident rate of 22 per 1000 workers is well in excess of the national rate for all industries of 14 per 1000 workers. Its fatality rate of 5.9 per 100 000 workers in 2009–10 was threefold that of the all-industry rate, which was 1.9 per 100 000 workers (Safe Work Australia 2011).

Role of design in accident prevention and control

Research has repeatedly confirmed that design is a significant cause of fatalities and severe injuries in construction. Gibb et al. (2004) in the UK, for example, meticulously studied 100 construction-accident cases and concluded that slight changes to the designs would have significantly minimised the risk in nearly half (47%) of the cases analysed. Similarly, an analysis of 450 construction-accident reports in the US found that in almost a third (30%) of cases, the hazards that caused the accidents could have been prevented by design changes (Behm 2005). In Australia, too, Creaser (2008) established that 37% of fatalities on construction sites involved design-related causes.

Traditionally, the designers' role has been to produce building designs compliant with the local building codes and the responsibility for site safety has been left to the builder. Nonetheless, research indicates that designers can significantly influence construction-site safety at concept and at preliminary and detailed design phases when they can make design choices that result in contractors having to make fewer site decisions. Accident Prevention through Design (PtD) derives from this notion. The approach of PtD is applied through the different phases of the design process to identify and mitigate – or even eliminate – hazards potentially encountered by workers during construction as well as maintenance of a building or structure. Hence, PtD is a very pertinent and needed solution/strategy to curb the distressing accident statistics in the construction industry.

PtD entails systematically identifying construction hazards and risks posed by a design and introducing an alternative design option that satisfies both the client's requirements and local building codes while ensuring a safe workplace for workers (Kamardeen 2010). It is one of the most effective means of preventing accidents as it eliminates hazards at source. Realising the critical role that design decisions have in creating safety risks and the urgent need to reduce the accident toll in the construction industry, many governments around the world have mandated accident PtD in their legislations. In the UK, for example, PtD has been mandatory since 1995 through the Construction Design and Management (CDM) Regulations. Similarly, the Australian government has mandated PtD in its latest Work Health and Safety legislation that took effect from January 2012. Designers are now required by law to include PtD in their design practices – a major modification to their previously accepted role.

Challenges facing designers for PtD adoption

Designers are faced with critical challenges for effective adoption of PtD in their practices because of the traditions that: (1) design and safety typically remain as separate knowledge domains (Lew and Lentz 2009); and (2) safety has been left for contractors to resolve on site. These traditions have triggered many designers to have little understanding of how design decisions impact on safety risks (Vasconcelos, Soeiro and Junior 2011).

In order to implement PtD, designers must thoroughly know the construction contexts of their design. Every design creates a construction context, which dictates the level of hazard resulting from the design choice. The construction context can be usefully conceived of as including five elements: site settings – site terrain condition, space and accessibility, road and traffic condition and vicinity; task settings – location of the task and temporary structures/facilities needed to build it; materials – type and nature of materials used; equipment – type and nature of equipment used; and labour – type and skill level. A construction accident can emanate from one, or from a combination, of these contextual elements. PtD must therefore perform a what-if analysis for all possible design options for a building project to take account of their construction contexts and make safe design choices, using an iterative process from inception through to the detailed design phases.

The critical challenge here is the skill gap of designers. Recent industry surveys showed that many designers do not have either this kind of knowledge or the skills to undertake PtD effectively (Zou, Yu and Sun 2009; Cooke, Lingard and Blisman 2008), thus continuing to expose construction workers to unnecessary risks and themselves to new legal liabilities. Since designers are predominantly trained in design principles and they do not often work on construction sites, they usually have only limited practical knowledge of construction operations and contexts triggered by their designs. Detailed knowledge of construction technology, workplace health and safety, and risk management is essential to undertake meaningful PtD analyses.

Hence, the availability of a mechanism to provide designers with integrated knowledge and decision support is of critical importance to overcome this challenge, thereby fostering PtD adoption in industry. Moreover, any new system proposed should be affordable and available for most designers regardless of their organisational size.

Mobile computing for PtD

Spiteri and Borg (2006) some time ago argued that designers constantly require *modular* and *just-in-time* decision support during the design process. *Just-in-time* implies that knowledge and information are provided immediately when they are required to ensure that correct decisions are taken at differing design stages. This knowledge has to be presented in a *modular* format, making it possible to structure the information required into components that can be translated into usable chunks, which can then be *reused elsewhere* in similar design situations. They further suggested that mobile computing technologies offer ideal tools for providing the right knowledge within the right context at the right time to designers.

Recent advances in mobile computing technologies provide unique opportunities for developing cost-effective, novel decision-support systems for the construction industry. The management of knowledge and information made possible by mobile computing devices (e.g. personal digital assistants (PDAs), smartphones and iPads), together with mobile apps, has critical aspects that best suit and appeal to designers as they can provide any-time, anywhere access to content, just-in-time support and a media-rich interactive environment (Acar et al. 2008). Many construction-specific mobile apps are already in use, ranging broadly from specific-purpose calculators to design, collaboration and site-monitoring tools (Mike 2011; ConstrucTech 2011). In contrast, and despite this proliferation, the potential of mobile computing technologies for accident PtD has not yet been explored.

Focus of the book

Research shows that falls, strikes by objects, and electrocution are the top-three agents of workplace accidents in the construction industry. These represent over 80% of accidents during the construction, maintenance and demolition of buildings and other structures (Ozanne-Smith et al. 2008; Bureau of Labor Statistics 2010). Out of these top-three agents, falls are a dominant cause of fatalities and serious injuries in construction globally. In the US and Hong Kong, for instance, almost 50% of construction fatalities are due to falls from heights (Bureau of Labor Statistics 2010; Chan et al. 2008). Likewise, Australian Safety and Compensation Council (ASCC) (2009) reported that falls from heights represent the largest cause (one-third) of all construction fatalities in Australia. Additionally, falls are found to be the most expensive workplace risk in several countries. Courtney et al. (2001), for instance, estimated the annual cost of fall-related workplace injuries in the US to be nearly $6 billion in the year 2000. Workplace injuries and illness in the Australian construction industry cost about $12.42 billion in 2008, of which falls accounted for a third, about $4.10 billion (Safe Work Australia 2010). Similar statistics can be found in many different countries. These allude to the standpoint that construction falls are creating a distressing economic burden globally. Prevention and control of falls are therefore a key priority in construction; hence, this book is focused on this agenda.

In light of the critical issues and knowledge gaps discussed above, this book aims to establish a new knowledge base and system models necessary for the development of a mobile computing solution that can assist designers in reducing the high toll of construction falls by PtD strategies. The aim of the book is achieved by realising the following three critical objectives:

1   creating a new PtD knowledge base to model the relationships between fall risks and design decisions;
2   developing a new PtD mobile app to assist building designers in fall PtD; and
3   evaluating the practical implications of the fall PtD mobile app for building designers and the construction industry as a whole.

Falls occur in all types of construction projects such as buildings, civil engineering structures and specialist trade services. Nonetheless, this research covers only buildings in order to make the study and its findings more pertinent and effective.

Research process

The research that underpinned the development of this book consisted of five facets, and each facet involved different research methods, as illustrated

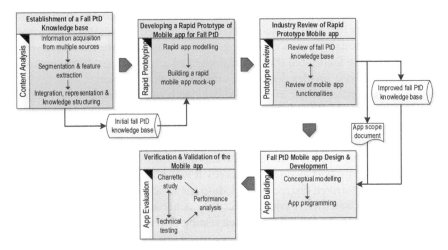

Figure 1.1 Research process.

in Figure 1.1, as appropriate to achieve the research objectives towards realising the final aim. The facets were:

- establishing a new knowledge base to support fall PtD;
- developing a rapid prototype mobile app for the purpose of reviewing the new knowledge base and for acquiring user expectations of app functionalities;
- industry review of rapid prototype mobile app;
- developing a working mobile app for fall PtD;
- verifying and validating the mobile app.

The establishment of a new fall PtD knowledge base was fundamental to the development of the mobile app. The knowledge base was built through a detailed content analysis of documents and literature such as research publications, workplace health and safety authority manuals and publications, as well as industry reports. The content analysis was totally qualitative in nature in that information from multiple sources was extracted and grouped under themes and sub-themes, and then it was integrated to form PtD suggestions under the themes and sub-themes. In parallel, 3D illustrations were also developed to strengthen the PtD suggestions with visual explanations.

It was essential to: (1) review the PtD knowledge base before using it in the app, and (2) acquire user expectations of the app functionalities to support fall PtD. A rapid prototype app/mock-up was developed based on the author's ideas of the functionalities required to support PtD. Following that, an industry review of the rapid prototype was administered in that the prototype was demonstrated to industry professionals who are involved in design reviews for safety. Their approvals of the PtD knowledge base and the app functionalities as well as suggestions for further improvement were acquired. These facilitated the establishment of a solid-scope document and a knowledge base for developing an effectively working app.

With the verified scope and the knowledge base, the research proceeded to develop a working mobile app for fall PtD. Conceptual models such as a knowledge representation map, decision logics, app layout model and app operation model were first created, which were then followed in programming the app in Apple's software development kit, Xcode.

Finally, the app was evaluated whereby it was both validated and verified. In a case-study-driven charrette workshop for validation, the app was cross-checked as to whether it could effectively achieve its intended purpose of harnessing fall PtD. Similarly, repeated testing of the app on multiple mobile devices such as iPhones and iPads was carried out to verify whether the app was programmatically complete and coherent. While the validation findings concluded that the app was beneficial for supporting fall PtD in the construction industry, some technical issues were detected in the app, which were then resolved at the fine-tuning phase.

Structure of the book

The book comprises seven coherent chapters to array discussions in a logical manner to enable easy reading, comprehension and knowledge building. The contents of each chapter are outlined below.

- This introductory chapter lays the foundation for cogent discussions in the subsequent chapters by explaining the key theme, focus and the research process that propelled the evolvement of the book.
- Chapter 2 elaborates on the paradigm of accident PtD. It explains the concept and the process of PtD along with their benefits and related regulations. It also discusses challenges facing the adoption of PtD and then identifies various strategies to overcome them.
- Chapter 3 explores design solutions for fall prevention in construction, which constitute the contents of a new fall PtD knowledge base developed in this research study. The chapter first provides an account of PtD solutions that can be applied at the early design stage. Then, PtD solutions pertinent to different building elements that might be incorporated at the detailed design stage are discussed.
- Chapter 4 expounds the development of a mobile app for fall Prevention through Design in construction. First, the stages, techniques and tools involved in producing mobile apps are explained in an easy-to-grasp style. Subsequently, these are operationalised in creating a new mobile app for fall PtD whereby the steps of app scope analysis, conceptual

designing and physical building of the app are detailed in the chapter. The important practical lessons learnt during the development process are also shared in the spirit of alerting future app developers.

- Chapter 5 discusses the evaluation of the new fall PtD app and the results. First, a framework of app evaluation criteria is developed to ensure an effective process. Then, the administration of a case-study-driven charrette workshop to evaluate the app in accordance with the framework is explained. Finally, the evaluation outcomes are compared and contrasted with the terms of reference.

- Chapter 6 demonstrates how PtD can effectively be incorporated into design and engineering curricula in tertiary institutions. Grounding on the paradigms of constructivist pedagogy and deep learning, a sample module design for PtD education is showcased along with model-assessment tasks.

- Chapter 7 concludes the book by highlighting the key findings in relation to the aim and objectives that were set forth in the introductory chapter. The chapter also explains the theoretical and practical implications of the key findings. Finally, it provides directions for future research.

The chapters in the book may be read in any sequence without difficulties in understanding the contents. Nonetheless, following the logical sequence laid out in the book will facilitate gradual knowledge building for readers who are new to the topics explored.

The mobile app developed in this study can be found in Apple store with the search keyword 'Fall PtD' and be installed on Apple devices such as iPhones and iPads for free.

References

Acar E, Wall J, McNamee F, Carney M and Öney-Yazici E. (2008) Innovative safety management training through e-learning. *Architectural Engineering and Design Management*, 4(3): 239–250.

Australian Safety and Compensation Council (ASCC). (2009) *Compendium of Workers' Compensation Statistics Australia 2006–07*. URL (accessed 9 Nov. 2009): http://www.safeworkaustralia.gov.au/NR/rdonlyres/6661279D-D04E-4142-97E7-69997CEC0157/0/PARTEJourneyclaims200607.pdf.

Behm M. (2005) Linking construction fatalities to the design for construction safety concept. *Safety Science*, 43(8): 589–611.

Bureau of Labor Statistics. (2010) Census of fatal occupational injuries summary 2009. *Economic News Release*. URL (accessed 10 Feb. 2012): www.bls.gov/iif/oshcfoi1.htm.

Chan A, Wong F, Chan D, Yam M, Kwok A, Lam E and Cheung E. (2008) Work at height fatalities in the repair, maintenance, alteration and addition works. *Journal of Construction Engineering and Management*, 134(7): 527–535.

ConstrucTech. (2011) Construction apps extend functionality. URL (accessed 9 Jan. 2012): http://www.constructech.com/news/articles/article.aspx?article_id=8864.

Cooke T, Lingard H and Blisman N. (2008) ToolSHeDTM: The development and evaluation of a decision support tool for health and safety in construction design. *Engineering, Construction and Architectural Management*, 15(4): 336–351.

Courtney T, Sorock G, Manning D, Collins J and Holbein-Jenny M. (2001) Occupational slip, trip, and fall-related injuries: Can the contribution of slipperiness be isolated? *Ergonomics*, 44(13): 1118–1137.

Creaser W. (2008) Prevention through Design (PtD) safe design from an Australian perspective. *Journal of Safety Research*, 39(2): 131–134.

Gibb A, Haslam R, Hide S and Gyi D. (2004) The role of design in accident causality. In: Hecker S, Gambatese J and Weinstein M (eds), *Designing for Safety and Health in Construction: Proceedings, Research and Practice Symposium*. Eugene, OR: UO Press.

Health and Safety Executive (HSE). (2010) *Statistics on Fatal Injuries in the Workplace 2009/10*. URL (accessed 10 Jan. 2012): http://www.hse.gov.uk/statistics/fatalinjuries.htm.

Hu K, Rahmandad H, Smith-Jackson T and Winchester W. (2011) Factors influencing the risk of falls in the construction industry: A review of the evidence. *Construction Management and Economics*, 25(3): 397–416.

Kamardeen I. (2010) 8D BIM modelling tool for accident prevention through design. In: Egbu C (ed.), *Proceedings of the 26th ARCOM Conference*. University of Leeds, UK, 6–8 Sept. 2010. Leeds: ARCOM. Vol.1, pp. 281–289.

Labour Force Survey. (2009) *Safety and Health for the Construction Industry*. URL (accessed 11 Oct. 2010): http://www.statistics.gov.uk/CCI/Nscl.asp?ID=5316&Pos=2&ColRank=1&Rank=80.

Lew J and Lentz T J. (2009) *Strategic Education Initiatives to Implement Prevention through Design (PtD) in Construction*. URL (accessed 26 Feb. 2013): http://ascpro0.ascweb.org/ archives/cd/2009/paper/CEGT79002009.pdf.

Mike N. (2011). *How Mobile Apps Changed the Construction Industry*. URL (accessed 9 Jan. 2012): http://www.onthegoware.com/mobile/mobile-apps-changed-construction-industry.

Ozanne-Smith J, Guy J, Kelly M and Clapperton A. (2008) *The Relationship Between Slips, Trips and Falls and the Design and Construction of Buildings*. URL (accessed 25 Sept. 2010): http://www.monash.edu.au/miri/research/reports/muarc281.pdf.

Safe Work Australia. (2010) *Compendium of Workers' Compensation Statistics Australia 2007–2008*. URL (accessed 28 Sept. 2010): http://www.safeworkaustralia.gov.au/NR/rdonlyres/56B40AF3-C5E8-4D7B-8CFAED91966DFE6F/0/Compendium200708.pdf.

Safe Work Australia. (2011) *Construction Fact Sheet*. URL (accessed 5 Jan. 2012): http://safeworkaustralia.gov.au/AboutSafeWorkAustralia/WhatWeDo/Publications/Pages/FS2010ConstructionInformationSheet.aspx.

Spiteri C and Borg J. (2006) Towards a conceptual framework for mobile knowledge management support. In: Marjanovic D (ed.), *Proceedings of 9th International Design Conference: Design 2006, Dubrovnik, Croatia, 15–18 May*. Dubrovnik, Croatia: Design Society, pp. 1259–1266.

Vasconcelos B M, Soeiro A A V and Junior B B. (2011) *Prevention through Design: Guidelines for Designers*. URL (accessed 26 Feb. 2013): http://repositorio-aberto.up.pt/bitstream/10216/55162/2/29026.pdf.

Zou P, Yu W and Sun A C S. (2009). An investigation of the viability of assessment of safety risks at design of building facilities in Australia. In: Lingard H, Cooke T and Turner M (eds), *Proceedings of the CIB W099 Conference*, Melbourne, Australia, 21–23 October 2009, Melbourne: RMIT University. CD-ROM, paper No 12.

2 Accident Prevention through Design in construction

Introduction

Accident prevention and control have presented a persistent global challenge for the construction industry, which has significantly higher rates of workplace fatalities and injuries than the average rates for all industries in many countries. Increasingly, research has established a strong correlation between design methods and workplace accidents in construction. Accident Prevention through Design (PtD) has therefore been proposed as an effective tactic for improving construction workers' safety. It is therefore of paramount importance that the concept and principles of accident PtD are understood clearly before embarking on the journey of implementing it in practice. Hence, this chapter aims to provide a detailed account on accident PtD. First, the origin, concept and process of PtD are expounded. Then, benefits of PtD for construction stakeholders are explained. After that, regulations surrounding PtD are described, followed by practical challenges facing PtD adoption. Next, strategies for improving PtD adoption are explored. Finally, conclusions are drawn showing the way forward.

The origin and concept of PtD

With construction being regarded as one of the most dangerous industries in many countries, workplace health and safety (WHS) management in the industry is of critical importance. In industrialised countries construction-site accidents contribute to 25 to 40% of workplace fatalities (Martínez Aires, Rubio Gámez and Gibb 2010). In addition to causing physical harm to human beings, construction accidents also delay the project's progress and add to construction costs significantly (Wang, Liu and Chou 2006). A plethora of research has been undertaken globally to date to identify and understand factors contributing to workplace injuries in a bid to better manage and improve workplace safety. Most of these have focused on construction-site as well as builder-related factors. The traditional design-bid-build procurement model also has led to the perception that ensuring the safety of construction workers is largely the responsibility of construction contractors (Hinze and Wiegand 1992; Rechnitzer 2001).

However, in recent years, especially in the last two decades, taking safety into consideration well ahead of the start of the construction work has gained momentum. The extensively cited research of Szymberski (1997) illustrated the relationship between a project schedule and the ability to influence safety, as shown in Figure 2.1. This seminal work puts forward the notion that risks inherent in a project are largely determined at the conceptual stage of the project, whereas when the project reaches the actual construction stage little can be done to improve safety. Since then, attention has been paid to establish the relationship between design and construction accidents. In this spirit, it has been concluded in many later research studies that incorporating the PtD concept into design stages can eliminate or significantly reduce workplace accidents. For instance, Behm's (2005b) analysis of 224 fatalities established that 42% of them could have been avoided if the PtD concept had been applied. A subsequent study using an expert panel further validated this conclusion (Gambatese, Behm and Rajendran 2008). Akin to this, a report on occupational accidents in the Australian construction industry outlined that design-related issues contributed to 37% of fatalities and at least 30% of injuries (Creaser 2008). In other words, 67% of workplace accidents could have been avoided through better designs in the Australian construction industry.

From the above discussion it can be argued that taking safety into account in the design stages is of vital importance in construction and hence the concept of PtD or design for construction safety has been formally proposed in the construction research agenda globally. Gambatese, Behm and Hinze

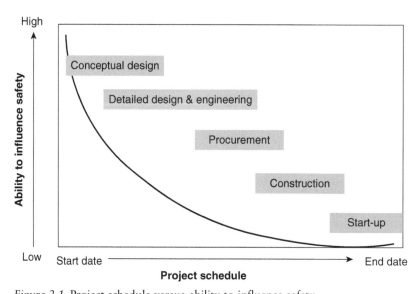

Figure 2.1 Project schedule versus ability to influence safety.

Source: Szymberski 1997, adapted by Weinstein, Gambatese and Hecker (2005).

(2005, p. 1029) defined: "designing for construction safety entails addressing the safety of construction workers in the design of permanent features of a project". This means PtD is proactive prevention of potential hazards and it is directed at eliminating potential hazards at source before their outbreak on site, that is, dealing with problems in the project design stages rather than in the project implementation stage (Fadier and De la Garza 2006).

Various terminologies and initialisms are being used by researchers and practitioners in different parts of the world to refer to the core concept of design for construction workers' safety. These include: Prevention through Design (PtD), Design-for-Safety (DfS), Safety through Design, Designing for Construction Safety (DfCS), Design and Planning for Construction Safety (DPfS), and Construction Hazard Prevention through Design (CHPtD). Essentially, all these different terminologies have the same idea. However, this book uses the terminology of Prevention through Design (PtD).

The process of PtD

PtD is an iterative process that is applied in different phases of design development, and involves a multidisciplinary team. Figure 2.2 illustrates the various phases in the design process and how PtD is integrated into these phases. It also shows the stages involved within the PtD process. PtD is not the responsibility of designers only, but a collaborative effort is also needed to achieve a better outcome. Thus, the PtD analysis team should be composed of experienced experts from different disciplines so that safety can be considered from multiple perspectives. This team may include the project manager, architect, engineers, construction manager, foremen, safety officer, subcontractors, etc. Akin to value management studies that are facilitated by the client's project manager, PtD studies could also be coordinated by this person. The benefit of this is that leverage by the client's project manager is likely to improve commitment from all stakeholders to the project. Moreover, specifying PtD as one of the expected services when drafting the service contract for these different stakeholders will improve its adoption in places where PtD is not a legal requirement. Detailed explanations on how PtD should be accomplished in the different design phases such as concept design, preliminary design and detailed design are provided below.

The design process begins by defining client's requirements. In this phase, the design team works closely with the client to determine the requirements for the project in terms of its functionality, performance level, design and aesthetic features, and other characteristics such as sustainability, budget and time constraints. The outcome of this phase is a client's brief, which serves as terms of reference for the subsequent phases. A noteworthy point is that the client's brief can be revisited and modified during the other phases in view of practicality and viability.

In the concept design phase, alternative project solutions to meet the client's requirements are identified. Following that, feasibility assessments of

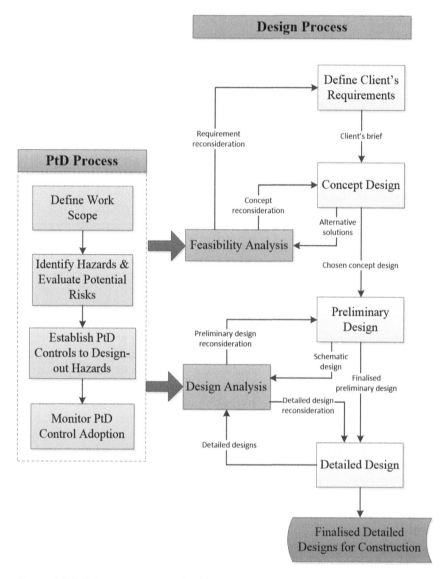

Figure 2.2 PtD integration into the design process.

the concept proposals are undertaken with the aim of selecting the optimum project solution that will be viable both financially and technically. Concept design PtD forms part of feasibility assessments. Its objectives are to evaluate concept design proposals from the perspective of workers' health and safety and to establish a safe design basis for a chosen design concept to carry on into the preliminary design phase. In this analysis, considerations need to be given to both the project site context and concept design alternatives. Variables considered for studying the project site context may include the following (Thorpe 2005, p. 52):

- location/vicinity of site – for example, desert, remote areas, highways, etc;
- previous site usage – for example, tip, factory, agriculture, etc;
- known residual/hazardous materials from previous usages/processes;
- underground services – electric, gas, water, sewerage, telecommunications, etc;
- streams, watercourses, ponds and other water features;
- water table level;
- results of soil or geological surveys;
- overhead cables – power and/or telephone wires;
- prevailing wind direction;
- adjacent buildings, their uses and types of foundation;
- restrictions on site access/egress – for example, width, height, times of use, etc;
- timing/season of construction;
- preserved features and ecological constraints;
- planning and/or building constraints.

By establishing the relevance of the above variables to the project, advance considerations and visibility to possible hazards and safety implications can be provided from the outset of the design. Accordingly, concept level PtD suggestions are drawn for a preferred concept design, outlining: preferred construction methods and techniques, orientation of the facility, types of temporary structures, types of materials and equipment, etc.

The chosen concept design is further developed in the preliminary design phase. An overall project configuration is defined whereby system and component level design requirements are established. Furthermore, schematic drawings, layout definition drawings and other engineering documentations are developed. PtD efforts in the preliminary design phase are an increment of the concept design PtD. First, an audit on the preliminary design is to be undertaken to ensure that concept design PtD suggestions have been absorbed adequately. Then, a hazard analysis will proceed from a facility level to a system and component level analysis. A study of project scope and sketch designs will help to identify and analyse different hazardous work trades to be involved in the project. These might include, for example, deep excavation, works at height, manual handling, handling of hazardous materials such as asbestos and waste, temporary works, erecting structures, crane use, works over water, confined-space works, etc. (Holt 2005; Thorpe 2005). Having identified the work trades and qualitatively analysed potential hazards in the project, high-level PtD suggestions will be made for risk control through preliminary design revision. Exhibit 2.1, for instance, illustrates how works at height may be analysed to derive high-level PtD suggestions. Once the preliminary design has been revised to take account of PtD suggestions, it then proceeds to the detailed design phase. The detailed design development should be monitored to ensure that the preliminary design PtD suggestions are fully followed.

Exhibit 2.1: Preliminary design PtD

What are the risks associated with works at heights?

> The main risk associated with works at heights is falling, which can lead to death or serious injuries.

Why do falls from heights occur?

- failure of height-access modes – i.e. ladders and scaffoldings;
- insufficient strength of structural elements, support members and platforms to bear the weight of workers (e.g roofing materials, skylights, formwork, etc.);
- insufficient, or lack of, edge protection;
- workers' movement beyond protected areas;
- failure of personal protective equipment – i.e. fall-arrest systems.

What can be done to eliminate the need for works at heights?

- consider off-site fabrication of elements that are constructed at heights;
- reduce the need for ladders and scaffolding by including hard platforms to allow the use of mobile access equipment such as hydraulic platforms, scissor lifts and boom lifts;
- design windows that can be fitted and maintained from inside the building;
- specify materials that require minimum maintenance;
- specify non-slip and strong surface materials and non-fragile roofing materials.

The detailed design phase produces different types of design drawings and specifications for components and systems (such as architectural, structural, mechanical, electrical and plumbing (MEP) and fire systems, etc.), carefully detailing their functions and interfaces so that they can be built on site. Detailed PtD design efforts analyse the designs and specifications of individual elements to locate hazards caused by design decisions. The risk-analysis step then determines which of the hazards are removable by design revisions and which are to be closely managed on site. Subsequently, PtD suggestion reports are to be produced for design revisions. The monitoring step oversees how the PtD suggestions are embraced in revising the original designs. Exhibit 2.2 shows, for example, a detailed design stage PtD analysis for a roof. Undertaking such a detailed PtD exercise entails comprehensive knowledge about material properties, construction procedures and temporary works. Brainstorming, focus-group discussions, checklists and hazard databases can be useful means for effective hazard identification and PtD suggestion derivations.

Exhibit 2.2: Detailed design PtD for roof

Design details:

Drawing reference: DWG 23 –V1
Design description: pitched roof for a two-storey house with roof tiles, skylights and fibreglass panels.

What are the hazards inherent in the design?

- slope of the roof;
- location and placement of skylights;
- low strength roof materials (fibreglass panels);
- works at heights on roof;
- ladder use for access to roof.

What are the potential risks and their severity level?

- falls from roof edges – high severity;
- falls through skylights and/or fibreglass panels – high severity;
- trip, slip and falls on the roof – moderate severity.

What design modification (PtD) measures can be undertaken?

- minimise the pitch of the roof to avoid workers slipping off the roof;
- design permanent anchorage points at 2.5 m intervals to provide tie-off points for fall-arrest systems during construction;
- place skylights on raised (200–300 mm high) curbs and away from roof edges;
- install permanent guardrails around skylights and around roof materials that are unsuitable for walking on (i.e. fibreglass panels);
- design-in location brackets for ladders to prevent slipping sideways;
- include clearly marked access and movement paths for workers.

Benefits of PtD

PtD promises many benefits for construction project parties. The most notable and direct benefits are a reduction in workplace accidents and improved workers' health (Toole and Gambatese 2008). Manuele (2008) highlighted several indirect benefits that can be obtained by adopting PtD, including: productivity improvement, operating cost reduction, avoidance of costly retrofitting, and overall risk minimisation. Specific modes that emanate from these benefits are as follows:

- proactive identification and avoidance of hazards at the design stage are considered safer and more cost-effective than reactive management of risks at the construction stage;
- increased safety achieved through PtD can minimise the total construction cost, resulting from reduced workers' compensation insurance costs for builders and improved productivity on site;
- project duration can be shortened or time overruns can be minimised as a consequence of effective PtD practices that lead to fewer injury-related delays;
- design professionals are more knowledgeable about technical issues, such as stresses and electricity in construction projects; therefore, the PtD concept of requiring this group to consider safety in their work can achieve better use of human resources;
- by involving most parties of the project in PtD discussions, it helps to raise safety awareness among the parties, especially that of the designers, in their efforts to reduce construction injuries; this has important implications for construction practices;
- construction safety awareness among project parties can be raised as PtD requires the involvement of all parties in discussions; this will have long-term implications for the industry in its efforts to reduce workplace accidents; and
- it strengthens the link between construction and the social and human resources' dimensions of sustainability; the social and human resources' dimensions of sustainability campaign for the principle that any project built must be socially harmless so that it will not cause injustice to any individual or group of people; PtD ensures that building designs do not harm the very workers who build them.

Furthermore, Ertas (2010) reported four case studies in which the business value of PtD was empirically analysed. Many positive impacts of PtD were demonstrated from the case studies, which include:

- design changes made in view of PtD caused reductions in the requirement for workers and supervisors to accomplish their respective construction tasks, resulting in significant savings in people costs;

- the need for costly personal protective equipment and engineering controls was eliminated;
- health-related absenteeism was reduced dramatically – workers became healthier, happier and more satisfied with the work;
- workers' morale was enriched, leading to better workmanship and quality of work;
- trust and confidence in the business among workers and workers' unions were enhanced;
- relationships between different management divisions in the workplace improved; and
- the reputation and esteem of the business increased among employees, clients and insurance providers.

In summary, the application of PtD in the design process not only safeguards workers who build the design on site, but also offers many business values to design and construction firms. Moreover, it reinforces the implementation of sustainability principles in its social and human resources' dimensions.

Workplace Health and Safety regulatory framework and PtD duty of care

Recognising the important contributions that good design can make to improving workplace health and safety, a substantial degree of 'duty of care' is assigned to designers through the Workplace Health and Safety (WHS) regulatory framework in many countries. Duty of care requires designers to do everything 'reasonably practicable' to protect the health and safety of workers by demonstrating that they have responded to hazards in the best way possible within the constraints of the work environment and have exercised due diligence. Failure to exercise due diligence that results in a breach of duty of care will be considered professional negligence, which may result in civil – or even criminal – legal action.

Figure 2.3 illustrates the pyramid of regulatory framework that elaborates the PtD duty of care of designers and their responsibilities. Acts and regulations are mandatory, and non-compliance is considered an offence and can result in a fine, issuance of an improvement or prohibition notice and/or imprisonment. Codes of practice, standards and best-practice manuals, on the other hand, are voluntary guidance materials. The extent of operationalisation of this framework in the context of PtD varies from country to country. For instance, while some countries, such as the UK, Australia and Singapore, have all the levels of the pyramid in place and have mandated PtD through their WHS Acts and relevant regulations, other countries have, to date, been facilitating its adoption only via voluntary approaches.

In 1995, the UK mandated PtD through its Construction Design and Management (CDM) Regulations in that avoiding foreseeable health and safety risks became a duty of designers. Other European Union (EU) nations

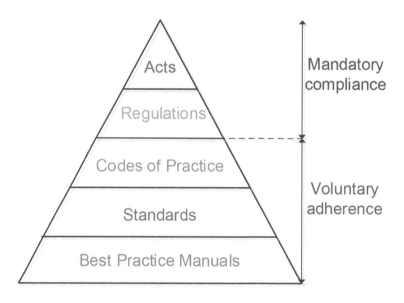

Figure 2.3 Regulatory pyramid.

Source: adapted from Australian Safety and Compensation Council 2006, p. 12

also mandated, or strongly encouraged, PtD. The Directive 89/391/EEC was adopted in 1989 to provide a legal framework to guide construction workers' and employers' practices to ensure worker safety (Martínez Aires, Rubio Gámez and Gibb 2010). In particular, this initiative required the creation of two coordinators – one for design, and the other for construction – to eliminate or reduce construction hazards before the start of construction work. A further assessment on the extent to which this initiative has achieved accident prevention in construction has shown that since the legislation was put into force, a significant majority of countries have experienced decreased incidence rates (Martínez Aires, Rubio Gámez and Gibb 2010). Since 1994, the Directive 92/57/EEC, which was incorporated into all EU countries' national legislations, specified that designers have a legal obligation to take account of construction safety in their work (Frijters and Swuste 2008).

In Australia, a regulatory framework for safe design was established nationwide under the National Occupational Health and Safety Commission (NOHSC) in the late 1980s (Creaser 2008). Following this initiative, the Safe Design Project was undertaken in 1998/1999 to examine design-related issues that impacted WHS, and then reports to review the existing situations of safe design were prepared. Moreover, 'eliminating hazards at the design stage' was uplifted to the status of 'one of the five national priorities' in the Australian National Occupational Health and Safety Strategy

2002–2012 (NOHSC 2005). Subsequently in 2006, the Australian Safety and Compensation Council produced 'Guidance on the Principles of Safe Design for Work' while the WorkCover (equivalent of HSE in the UK and OSHA in the US) introduced the Construction Hazard Assessment Implication Review (CHAIR) (Mroszczyk 2008). Likewise, the Federal Safety Commissioner introduced 'Safety Principles and Guidance' that contained eight safety principles for improving WHS in the construction industry. The fourth principle stated: '*industry participants should ensure that safe design and constructability is considered at the planning and procurement stages to reduce or eliminate hazards and control risks before construction commences*' (Office of the Federal Safety Commissioner 2008). While the previous regulatory framework largely consisted of codes of practice and industry standards, which meant that their enforcement was voluntary, the latest Work Health and Safety Act 2011 has raised design for safety to the level of a national law (NSW legislation 2011). The Australian government has mandated the use of PtD in its new Work Health and Safety (WHS) legislation that took effect from 2012. Designers are now required by law to include PtD in their design practice – a major modification to their previous responsibilities.

In the US, the need of PtD for curtailing accidents was recommended in the National Safety Council's 1955 *Accident Prevention Manual*. Nonetheless, it did not materialise until the 1990s. In the early 1990s, the benefits of PtD were slowly recognised and thus started gaining momentum. Following that, in 1995, the National Safety Council established the Institute for Safety through Design with a mission to integrate PtD into design practice. The institute accomplished a good deal by facilitating seminars, workshops, symposia and conferences for practising engineers and academia. Since 2005, Occupational Safety and Health Administration (OSHA) have been convening a course entitled 'Design for Safety' via the Design for Construction Safety Workgroup, which comprises ten professional institutions. Similarly, the National Institute for Occupational Safety and Health (NIOSH) facilitated PtD workshops for industry sectors in 2007 and 2011. Moreover, PtD has become one of the ten focus areas of the National Occupational Research Agenda (NORA) Construction Sector Council since 2006 (Prevention through Design 2013). On this track, January 2012 saw a rapid advance in that a national standard on PtD came into effect, namely, 'Prevention through Design: Guidelines for Addressing Occupational Hazards and Risks in Design and Redesign Processes'.

In brief, governments are able to use a range of tools to change industry practices with regard to PtD. These tools include: legislations; regulations; codes of practice; standards; guidance and information packages; and awareness and educational campaigns. While legislations mandate PtD and hold designers legally accountable for non-compliance, other means promote its adoption voluntarily. Realising the effectiveness of PtD to eliminate hazards at source and the urgency to reduce workplace incidents, many governments around the world have adopted all the tools while others have adopted only some of them.

Challenges for PtD adoption in industry

Even though the benefits of PtD have been well recognised and endorsed by legislations and government official documents as well as by industry practitioners, successful PtD adoption in industry still faces significant challenges. These include: PtD competency gap among design professionals; legislative hiccups; procurement and contract issues; and financial considerations.

While legislations in many countries have established the obligation of the designer to reduce safety hazards, unfamiliarity with the implementation of safety measures in design is always blamed for failing to comply with these regulations (Frijters and Swuste 2008). A lack of education in this area has been identified as having prevented designers from applying PtD in their work (Gambetese 2003; Toole 2005). Most designers do not graduate with formal training in PtD or construction safety, nor do most design curricula include WHS or PtD (Zarges and Giles 2008).

In countries where PtD has not yet been mandated through legislation, problems with other aspects of the legislative system may discourage designers to actively take on the responsibility of risk reduction in their work. This has been particularly highlighted in the US, where designers are generally advised not to be concerned with safety in their design to avoid any possible liabilities when an accident occurs (Hinze and Wiegand 1992; Gambatese, Hinze and Haas 1997). Moreover, even though mentioned by OSHA Acts, designers are actually absolved from taking on the legal responsibility for the safety of workers (Hallowell 2011). Deficiencies in the existing legislative framework also lead some researchers to comment that diffusion of PtD in the US is lagging behind other countries such as Australia and the UK (Toole and Gambatese 2008).

The design-bid-build model for project delivery can inevitably discourage the application of PtD from the perspective of financial considerations (Hallowell 2011). PtD may not be adopted voluntarily in the design-bid-build procurement model, unless extra funding is set aside. This is particularly problematic as increased costs associated with the adoption of PtD are identified as a major concern by construction professionals who participated in the research of Gambatese, Behm and Hinze (2005). Furthermore, when drawing up construction contracts, the design-bid-build model actually excludes the responsibility of designers in ensuring the safety of temporary occupants or, more specifically, construction workers (Gambatese 1998; Behm 2004; Behm 2005a).

PtD is perceived to increase the construction cost, and current procurement contracts and schedules do not include PtD as one of the items that consumes time and cost (Zarges and Giles 2008). However, in reality, individual elements of PtD may cost more, but the total cost will be lower, and savings will be obvious when it is looked at as a whole. Likewise, effort and time will need to be invested up front, but net savings will be realised later to compensate for that, which result from a reduction in delays caused by incidents.

To sum up, the challenges for PtD adoption arise from different perspectives and may be legitimate. Nonetheless, these challenges can no longer be quoted as excuses for not adopting PtD, because PtD adoption is critical to the well-being of workers as well as constituting a moral – if not legal – obligation of designers. The challenges suggest that there is a need for integrated and concerted efforts from legislative authorities, educational institutions, research communities and the construction industry entities (i.e. project managers, designers, safety specialists, procurement specialists, subcontractors and material manufacturers) for developing practical strategies to facilitate effective PtD adoption.

Strategies for facilitating PtD adoption

Research efforts around the world have been undertaken to identify prudent and practical ways to overcome the PtD adoption challenges. Findings of these studies suggest three broad categories of strategies, known as: leveraging research and education; influencing industry standards and practice; and initiating policy changes.

Manuele (2008) suggested three capacity building strategies to foster PtD adoption at a larger scale, which are: expanding the existing body of knowledge around PtD through research; integrating PtD course materials into design and engineering curricula in tertiary education institutions; and establishing connections between academic institutions and researchers and the industry and labour unions to increase awareness. Similarly, Schulte et al. (2008) proposed a four-faceted framework to promote PtD adoption in industry, which comprises functional areas such as research, practice, education, and policy.

1   *Research* facet focuses on the advancement of the PtD knowledge base and diffusion of new knowledge to the academic community and practitioners.
2   *Practice* facet deals with influencing businesses or clients to demand safe designs, encouraging the design professional community to constantly increase their PtD awareness, and working collaboratively with WHS professionals in the design phases.
3   *Education* facet suggests augmentation of design and engineering curricula to incorporate PtD, as well as motivating professional institutions to include PtD elements in their competency-assessment frameworks that are used in degree accreditation programmes.
4   *Policy* facet recommends improving regulations, operating procedures, codes of practice and standards by incorporating PtD among the best practices.

Lin (2008) held a similar position with regard to clients demanding PtD-embedded designer services, but suggested more strategies concerning small businesses, industry culture and diffusion of information. Small businesses form a significant proportion of the overall industry, yet their inclination to adopt PtD is relatively low, which is often influenced by the fact that their profit margin is slim, which hampers investment in any new ventures. Hence, this section of the industry should be supported by providing simple, straight-forward and cost-effective solutions for PtD, for example, simple-solution booklets, positive case studies, newsletters and/or industry fact sheets. Moreover, training programmes should be geared toward small-scale businesses, and incentives should be offered for participation. Changing industry culture and mindset is crucial so that PtD becomes entrenched in organisational practices. Business leaders should cooperate in endeavours that increase PtD awareness, disseminate information about best practices, publicise business cases that show success stories, and provide information of cost savings attained through PtD. Toolkits and learning materials supporting the adoption of PtD should be developed and disseminated. For instance, a web-based database of best practices for safe design could be developed and made available; continual professional development (CPD) programmes on PtD might be offered by professional institutions; project case studies that demonstrate added business benefits and other tangible returns from PtD adoption could be created and disseminated by industry groups.

In a comprehensive fashion, Gambatese (2011) recommended five stones for establishing a solid foundation for successful PtD adoption, which are: culture; risk; organisation and project structure; physical form and function; and resources, tools and processes. Making a cultural and industry standard change is considered a cornerstone to PtD adoption. Five tactics were suggested to achieve this in practice, including:

- incorporate PtD into design and engineering curricula;
- recognise the value of designers for construction safety;
- establish clear responsibilities for all parties involved in safety;
- institute disciplinary collaboration for safety;
- integrate PtD with sustainability.

Equally important is a well-defined risk and liability-management framework. Existence of a clear mechanism for managing risks and liabilities that may result from PtD implementation would encourage designers to practise it willingly. In this vein, four risk and liability-management mechanisms were suggested by Gambatese (2011), which are:

- establish model contracts for design and construction, incorporating PtD services;
- clearly define the designer's PtD responsibilities and liabilities;

- introduce insurance policies that cover PtD liabilities;
- educate and train designers on construction hazards and risks.

Likewise, formal contractual arrangements can play a major role in obliging team members to commit to PtD adoption; accordingly, four avenues were highlighted by Gambatese (2011) to modify the current procurement structure, namely:

- advocate integrated project delivery;
- involve project clients in PtD discussions;
- define the designer's PtD roles;
- modify duties of the construction engineer.

Moreover, four recommendations were made by Gambatese (2011) to support PtD adoption by ensuring the availability of physical resources, namely:

- develop and disseminate examples of safe designs for construction elements;
- encourage product modularisation;
- promote the use of prefabricated components;
- produce automated construction technologies.

Finally, in terms of underpinning PtD adoption with soft infrastructure, six considerations were put forward by Gambatese (2011):

- introduce model PtD review processes and toolkits;
- embed PtD into design codes;
- develop design risk-assessment systems/methods;
- undertake and disseminate exemplary PtD case studies;
- create PtD resources and materials for design and engineering education and training;
- integrate PtD into professional licensing assessment and registration.

All things considered, the theme of 'PtD body of knowledge' was repeatedly highlighted by many researchers as an essential ingredient for facilitating PtD adoption. In this vein, it can be recommended that the development and dissemination of the following could make a significant impact on, and contribution to, rapid and effective PtD adoption in the industry:

- model PtD review procedures;
- design risk analysis and rating methods;
- safe design solutions and methods;
- exemplary PtD adoption case studies.

Conclusion

PtD is recognised as an effective way for eliminating or reducing hazards on construction sites, which cause workplace fatalities, injuries and health problems. Because of its well-established benefits for workers as well as businesses, many governments, research and education institutions, and industry bodies around the world have been promoting PtD through various mechanisms. Nonetheless, it is evident that PtD adoption is hitherto faced with many challenges. A key challenge is the still-developing body of PtD knowledge, which significantly impacts on the PtD competency of designers. Developing knowledge bases of safe design solutions for different design elements as well as toolkits that allow methodological analyses of designs is of paramount importance for overcoming this critical bottleneck. The next chapter establishes the contents for such a knowledge base, with a specific focus on fall Prevention through Design. The subsequent chapters of the book discuss the development of a PtD analysis support toolkit that utilises the new knowledge base.

References

Australian Safety and Compensation Council (ASCC). (2006) *Safe Design for Engineering Students: An Educational Resource for Undergraduate Students.* Canberra: Australian Safety and Compensation Council.

Behm M. (2004) *Establishing the Link between Construction Fatalities and Disabling Injuries and the Design for Construction Safety Concept.* PhD Dissertation. Oregon State University.

Behm M. (2005a) *Design for Construction Safety: An Introduction, Implementation Techniques, and Research Summary.* URL (accessed 27 Aug. 2014): https://www.onepetro.org/download/conference-paper/ASSE-05-724?id=conference-paper%2FASSE-05-724.

Behm M. (2005b) Linking construction fatalities to the design for construction safety concept. *Safety Science,* 43(8): 589–611.

Creaser W. (2008) Prevention through Design (PtD) safe design from an australian perspective. *Journal of Safety Research,* 39(2): 131–134.

Ertas A. (2010) Transdisciplinarity: Design, process and sustainability. *Transdisciplinary Journal of Engineering and Science,* 1(1): 30–48.

Fadier E and De la Garza C. (2006) Safety design: Towards a new philosophy. *Safety Science,* 44(1): 55–73.

Frijters A and Swuste P. (2008) Safety assessment in design and preparation phase. *Safety Science,* 46(2): 272–281.

Gambatese J A. (1998) Liability in designing for construction worker safety. *Journal of Architectural Engineering,* 4(3): 107–112.

Gambatese J A. (2003) Safety emphasis in university engineering and construction programs. *Special Issue on Construction Safety Education and Training: A Global Perspective,* Hinze J (ed.), *International e-Journal of Construction,* University of Florida, M.E. Rinker School of Building Construction, May 2003.

Gambatese J A. (2011) *Findings from the Overall PtD in UK Study and Their Application to the US. Prevention through Design: A New Way of Doing Business: Report on the National Initiatives.* URL (accessed 26 Apr. 2013): http://www.asse.org/professionalaffairs_new/PtD/Research%20Issues/John%20Gambatese.pdf.

Gambatese J A, Behm M and Hinze J. (2005) Viability of designing for construction worker safety. *Journal of Construction Engineering and Management,* 131(9): 1029–1036.

Gambatese J A, Behm M and Rajendran S. (2008) Design's role in construction accident causality and prevention: Perspectives from an expert panel. *Safety Science,* 46(4): 675–691.

Gambatese J A, Hinze J and Haas C. (1997) Tool to design for construction worker safety. *Journal of Architectural Engineering,* 3(1): 32–41.

Hallowell M. (2011) Prevention through Design tool for high performance sustainable buildings. In: *Proceedings of the 2011 Working Commission on Safety and Health on Construction Sites Annual Conference,* Washington DC, 24–26 August 2011.

Hinze J and Wiegand F. (1992) Role of designers in construction worker safety. *Journal of Construction Engineering and Management,* 118(4): 677–684.

Holt A. (2005) *Principles of Construction Safety.* Exford: Blackwell Science Ltd.

Lin M. (2008) Practice issues in Prevention through Design. *Journal of Safety Research,* 39(2): 157–159.

Manuele F. (2008) Prevention through Design (PtD): History and future. *Journal of Safety Research,* 39(2): 127–130.

Martínez Aires D, Rubio Gámez C and Gibb A. (2010) Prevention through Design: The effect of European directives on construction workplace accidents. *Safety Science,* 48(2): 248–258.

Mroszczyk J W. (2008) *Designing for Construction Worker Safety.* URL (accessed 24 Apr. 2010): www.asse.org/membership/docs/John%20Mroszczyk%20Article.doc.

National Occupational Health and Safety Commission (NOHSC). (2005) *National Standard for Construction Work.* Australia: NOHSC.

NSW legislation. (2011) *Work Health and Safety Act 2011.* URL (accessed 28 Aug. 2014): http://www.legislation.nsw.gov.au/maintop/view/inforce/act+10+2011+cd+0+N.

Office of the Federal Safety Commissioner. (2008) *Fact Sheet: Federal Safety Commissioner's Safety Principles.* URL (accessed 4 Feb. 2013): http://www.fsc.gov.au/sites/FSC/Resources/AZ/Documents/FSCsSafetyPrinciplesNewformat.pdf.

Prevention through Design. (2013) *History of PtD.* URL (accessed 26 Apr. 2013): http://www.designforconstructionsafety.org/challenges.shtml.

Rechnitzer G. (2001) *The Role of Design in Occupational Health and Safety: A Discussion Paper.* Melbourne: Safety Institute of Australia.

Schulte P A, Rinehart R, Okun A, Geraci C L and Heidel D S. (2008) National Prevention through Design (PtD) initiatives. *Journal of Safety Research,* 39(2): 115–121.

Szymberski R T. (1997) Construction project safety planning. *Tappi Journal,* 80(11): 69–74.

Thorpe B. (2005) *Health and Safety in Construction Design.* London: Gower Publishing Limited.

Toole T M. (2005) Increasing engineers' role in construction safety: Opportunities and barriers. *Journal of Professional Issues in Engineering Education and Practice*, 131(3): 199–207.

Toole T M and Gambatese J. (2008) The trajectories of Prevention through Design in construction. *Journal of Safety Research*, 39(2): 225–230.

Wang W, Liu J and Chou S. (2006) Simulation-based safety evaluation model integrated with network schedule. *Automation in Construction*, 15(3): 341–354.

Weinstein M, Gambatese J and Hecker S. (2005) Can design improve construction safety? Assessing the impact of a collaborative safety-in-design process. *Journal of Construction Engineering and Management*, 131(10): 1125–1134.

Zarges T and Giles B. (2008) Prevention through Design (PtD). *Journal of Safety Research*, 39(2): 123–126.

3 Safe design solutions for fall prevention in construction

Introduction

This chapter aims to establish a knowledge base of safe design solutions for fall prevention in construction projects. Fall Prevention through Design (PtD) first entails the identification of fall hazards caused by design methods. Safe Work Australia (2011) suggested scrutinising certain specific locations, elements and issues in designs for a project to identify fall hazards, namely:

- surfaces – stability, fragility, brittleness, potential to slip, strength to support load, slope and evenness;
- levels – working above or below the ground level;
- structures – stability of temporary and permanent structures;
- edges – open edges of floors, working platforms, walkways, walls and roofs;
- holes, openings or excavations that require guarding;
- work area entry and exit points.

Falls are also closely related to specific types of work. For instance, Huang and Hinze's (2003) study showed that roofing is the single most-frequent type of task performed when fall accidents occurred; this is followed by erecting structural steel and exterior carpentry as being works often associated with falls. Another research also demonstrated that falling from roofs is the greatest cause of fatalities (Dong et al. 2013). This suggests that PtD implementation for reducing the risk of falls should take into account a range of factors.

This chapter traverses through key elements in a building design and scrutinises the critical locations, identified above, in those elements. Then, safe design suggestions for fall prevention are presented under two categories, namely: early design-stage suggestions and detailed design-stage suggestions. As indicated in the first chapter, the safe design suggestions were first established through a detailed content analysis of documents and literature, such as research publications, workplace health and safety authority publications, and industry reports. Following that, the suggestions were refined through an industry review process. This chapter contains the refined suggestions; however, citations are also provided throughout the chapter to the literature sources.

Fall prevention at the early design stage

The nature of the project site and the scope of work dictate the level of risk involved in a construction project. Analysing the concept and schematic designs of a project will help to identify the layout of the building, the different work trades involved in a project and the risks posed by these. Based on the analysis, high-level design suggestions might be derived to minimise accidents. On this track, the following factors should be considered at the early design stage to minimise potential risks:

- building orientation to suit site conditions;
- building layout and morphology;
- safe construction methods;
- safe materials and equipment.

In the context of fall prevention at the early design stage, high-level design considerations are suggested below under appropriate subheadings, which were elicited from Thorpe (2005), Ghule (2008), WorkSafe Victoria (2008), WorkCover NSW (2009), Construction Industry Institute (CII) (2010), Speegle (2011), Safe Work Australia (2011), Safe Work Australia (2012), and Australian Building Codes Board (2008), and refined through industry reviews.

Project site

Characteristics of the project site may present fall hazards during construction, occupancy and maintenance of buildings. Obtain adequate information regarding the following ten factors and consider how safe your design is to build, occupy and maintain in view of these factors:

- severe weather conditions inherent to the project locality (e.g. high wind, snowfall, floods, etc.);
- prevailing wind directions;
- adjacent buildings (e.g. types of building, uses and types of foundation);
- limitations on site access/egress during construction and maintenance (e.g. width, height, load-bearing capacity and times of use);
- preserved features and ecological constraints;
- water bodies on site (e.g. ponds, streams, water features, etc.);
- overhead power cables and telephone lines;
- underground services (e.g. electricity, gas, sewerage, telecommunications, etc.);
- demolition requirements;
- residuals from previous site use.

These factors may warrant additional/special design requirements to ensure safe construction, occupancy and maintenance. Early coordination between the architect, owner, engineer, builder, etc. would help to derive pertinent safe design solutions for the project.

Extensions and renovations

Extensions and renovations of existing buildings pose numerous slip, trip and fall hazards to workers, which often result in serious lifelong suffering. Consider the below-described recommendations in designing for extensions or renovations.

1   Evaluate the structural stability/capability of members/parts of the existing building before adding further loads to them with extensions/ renovations. Early coordination between the architect and the structural engineer would be helpful. Additional loads may cause the structural members to fail and collapse, leading to serious falls and other accidents.

2   Design extensions and renovations such that the need for demolishing parts of the existing building is eliminated or minimised. Demolitions can cause serious risks to workers due to premature collapse of the structure being demolished and/or falls from demolished parts.

3   If demolition is essential for extension/renovation, investigate the project well and obtain the below-listed information to assist your design process (collaboration between the architect, owner, structural engineer, builder, etc. is recommended for this purpose):

    • the construction details of the parts to be demolished, including the materials, their strengths, the presence of cantilevered elements, holes, open edges, and any other general weaknesses;
    • the load-carrying capacity of adjoining parts of the building or land;
    • the need for possible temporary support structures for the part of the building to be demolished, and for adjoining parts and buildings.

4   Site new building elements/parts such that there will be minimum interactions with existing foundations and low-rise elements, such as steps, platforms and slopes, which are conducive to slips, trips and falls.

Sloped terrain

Working on steep slopes exposes workers to high risks as they could fall through the edges of slopes. Moreover, construction plant may lose balance and tip over. Even during the occupancy, buildings on slopes pose many slip, trip and fall hazards for users and maintenance workers. To minimise such risks, consider orienting the project layout so that the amount of work on steep slopes is reduced.

If work on steep slopes is unavoidable, provide warning and information about the site conditions in the design and construction documents to enable the builder to prepare for safe construction. Furthermore, design-in permanent signs, warning and barriers in such buildings where slip, trip and fall hazards for users and maintenance workers are present. Moreover, schedule footpaths, sidewalks and roadways around sloped work areas to be built early in the stage to provide stable footings for scaffolds, ladders and equipment.

Complicated plan shapes

The presence of multiple recesses, offsets or wavy sections in a building plan shape creates additional fall hazards during construction and maintenance because:

- setting up scaffolds around the building perimeter becomes difficult and this process itself presents fall hazards;
- accessing windows in recesses, corners and envelopes with curves for cleaning and maintenance is difficult and poses additional fall hazards.

Consider the following design suggestions in order to minimise these risks:

- reduce the number of recesses and offsets, or combine small recesses and offsets, to make them as large as possible;
- maintain consistent, large-sized recesses and offsets within a building.

If small recesses, offsets and non-standard shapes are unavoidable, contemplate the following design features:

- minimise the placement of windows and other elements that require regular cleaning and maintenance in difficult-to-access locations, corners and shapes of envelopes; alternatively, install windows that can be maintained from inside the building;
- use prefabricated walling and cladding that can be assembled with minimal work at heights;
- step floor plans and design-in aesthetic features, sitting areas, balconies, skyrise greenery, etc. on the ledges (see, for example, Figure 3.1); this will help break the total height and provide intermediate permanent platforms for construction and maintenance work; moreover, in the event of a fall, the fall distance will be limited to one floor, reducing the impact.

High-rise buildings

The main risk associated with the construction and maintenance of high-rise buildings is falls, leading to deaths or permanent disabilities. The following design measures are recommended to reduce the risk as far as possible:

Figure 3.1 Stepping floor plans in complicated shapes.

(Image courtesy of Benson Lim 2014.)

- specify prefabricated components and elements for parts that involve work at heights, for example, columns, beams, slabs, staircases, roofs, external walls, cladding, windows and other exterior work; prefabricated components and elements lend themselves to easier assembly and require less work at heights;
- design-in permanent anchors in appropriate locations on the prefabricated members along the perimeter (e.g. beams and columns) to provide stable connection points for fall-arrest systems during the assembly of building envelopes and their maintenance;
- design windows that can be cleaned and maintained from inside the building, without having to access them externally;
- choose materials that do not require frequent maintenance, such as repainting, for external envelopes of high-rise buildings.

Tapered buildings

It is harder to fix tapered/stepped scaffolds on the exterior of a tapered building during construction. As a result, scaffold stability, and thereby the safety of workers at heights, is often threatened. Design-in strong scaffold connection points and sockets along the perimeter of the tapered building to brace and continue scaffolds to reach the required height safely. Coordinate with the builder to identify suitable locations for such sockets and connection points.

When reflective glass is used on the walls of a tapered building, sun rays reflected from the glass surface can cause glare to motorists and other road users on nearby roads and the surroundings of the building. Hence, design a safe elevation for the building to avoid potential risks, taking into account the different positions of the sun on the horizon in winter and summer. Moreover, consider using non-reflective glass panels for tapered buildings.

Cantilevered buildings

Constructing a cantilevered building, as in Figure 3.2, requires sophisticated scaffolding systems and constant work at heights on temporary platforms, which poses serious fall risks for workers. Moreover, during the use of the building, it is riskier to reach facades, windows and soffits for cleaning and maintenance work.

Consider the use of precast elements to build cantilevered buildings as these elements will eliminate the need for temporary work platforms

Figure 3.2 Cantilevered building.

(Image copyright purchased from iStock and adapted.)

(scaffolds and formwork), reinforcement and concreting at cantilevered heights. The magnitude of work at heights is significantly reduced as the elements are cast on the ground and fixed in place by craning with minimum work at heights. Also, design-in an adequate number of anchor points on the precast elements to connect fall-arrest systems during construction and maintenance.

To facilitate safe maintenance of cantilevered buildings, consider limiting the heights of cantilevered parts to that which can be reached by a cherry picker, or similar equipment from the ground or a safe platform.

In-situ versus prefabricated construction

In-situ concrete construction presents numerous fall hazards for workers due to the magnitude of activities involved in formwork installation, reinforcement fixing and concreting. In order to minimise these risks, consider prefabricated elements and members for frames, roofs, walls, etc., which offset the requirement for formwork, reinforcement and concreting at heights, rather requiring only to be assembled.

If in-situ concreting is unavoidable, consider the following design suggestions to reduce fall hazards as far as possible:

- create dimensions and shapes of structural elements so that prefabricated formwork systems can be utilised rather than the traditional system that requires on-site assembly;
- wherever possible, use prefabricated reinforcement cages for columns and beams, and welded wire meshes for slabs and walls, rather than steel rods that require on-site assembly;
- specify shotcrete instead of poured concrete; by the use of shotcrete machines, concrete can be sprayed in to slopes, vertical walls and overhead, without workers having to reach these hazardous locations.

Material selection

In order to reduce fall hazards through effective material selection:

- specify building materials that do not require regular maintenance;
- state that materials and equipment are to be painted and/or insulated on the ground prior to installation;
- for safe handling, specify smaller-sized, lightweight materials and equipment for work at heights.

Fall prevention at the detailed design stage

While there are many contextual factors that may add to the risk of falls from heights for a construction project, not all of these can be addressed through detailed design measures. As a general principle of designing for safety, it is suggested that when risks cannot be designed out, construction safety can be improved through the 'design-in' strategy. This means that, for instance, when the project involves more pitched-roofing work, designers can specify the use of non-fragile roofing materials and mark access and movement paths clearly to reduce the likelihood of falls from roofs (Thorpe 2005, p. 43). Therefore, this section discusses specific design measures pertinent to the structure and layout of a building to counter fall risks. In other words, the focus is on the physical object itself, rather than administrative controls or training and educational programmes. The discussion is arranged according to the work elements and/or locations involved in construction works.

Foundations

Open excavations, footing locations and reinforcing steels can create trip and fall hazards for construction workers during the construction of foundations. The below-listed safe design solutions are suggested to reduce the incidence of falls during foundation works, which were elicited from Speegle (2011), CII (2010), and WorkCover NSW (2009), and then refined through industry reviews.

1   Open excavations present fall hazards for workers. Designing few large footings, rather than many small footings, would help to minimise the number of footing excavations required and thereby hazards. However, consideration should also be given to possible changes required for shoring methods and resultant hazards when the excavation size is changed.
2   Specify precast foundations as much as possible to eliminate shoring and reinforcement work in excavations and thereby fall hazards.
3   If in-situ foundations are unavoidable:

- dimension footings so as to maximise the use of commercial shoring/formwork systems for footings, which can help reduce hazards that workers are normally exposed to in conventional shoring;
- in order to provide a stable walking surface and to reduce trip falls during and before concreting, design the top reinforcement layers of spread and combined footings as well as those of raft foundations to be spaced at not more than 150 mm intervals in either direction (see Figure 3.3 for illustration).

4   Whenever slab-on-grade is designed:

- design surface-mounted footings (integrated foundations), rather than strip footings that require excavation work;
- specify prefabricated welded wire meshes for slab-on-grade to reduce hazards associated with on-site reinforcement work.

5   In extension/renovation projects, place new footings at a reasonable distance away from existing footings so that trip hazards can be minimised.

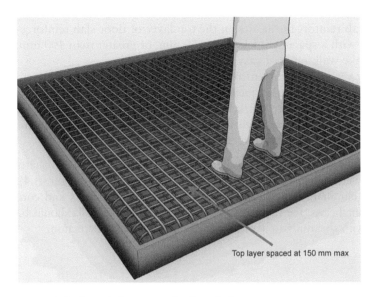

Top layer spaced at 150 mm max

Figure 3.3 Top reinforcement for foundations.

Reinforcement

Falls from elevation during reinforcement work could occur through unprotected edges and openings in upper floors while trip-related falls could occur on reinforcement and due to loss of balance during the manual handling of steel rods. Safe design solutions for reinforcement work, which were elicited from Thorpe (2005), WorkSafe Victoria (2008), Ghule (2008), CII (2010), OSHA Alliance Program Construction Roundtable (2010[?]a), and Speegle (2011), and refined through industry reviews, are presented in this section.

1   Fixing reinforcements at heights presents numerous fall hazards for workers, for example, tripping over steel bars and losing balance and falling when handling bars. In order to minimise such hazards, consider:

   - prefabricated reinforcement cages for columns and beams, which can be lifted and put in place just before concreting;
   - welded wire fabrics for upper floors and roof slabs, which allow rebar work in large sections rather than placing several individual steel bars for a single section.

2   If reinforcement-bar fixing is required at elevated floor levels:

   - specify small-sized reinforcing bars because large bars are heavy to handle and can cause trip and fall hazards;
   - for reinforcing core walls and lift shafts at heights, place horizontal bars on the outside of the vertical bars to eliminate risks caused by loss of balance when placing and threading the rebar into position.

3   For curtailing trip and fall hazards during reinforcement work:

   - for slab reinforcement, design the top layer of floor slab reinforcement with a spacing of less than 150 mm, or greater than 400 mm, to make walking on it easy;
   - allow vertical reinforcement bars for walls and columns to extend 1.8 m above the subsequent floor level, rather than using dowels and a vertical bar to mark the continuation of the reinforcement; this will remove trip hazards and the possibility of other serious injuries suffered at reinforcement extension points.

4   Continue reinforcement bars through openings on upper floors, lift shafts and core walls, which can be cut and removed after work (see Figure 3.4). This will eliminate falls through openings in slabs, lift shafts and core walls. Nonetheless, the reinforced openings in slabs and walls should be covered with sheathing to avoid other hazards during construction.

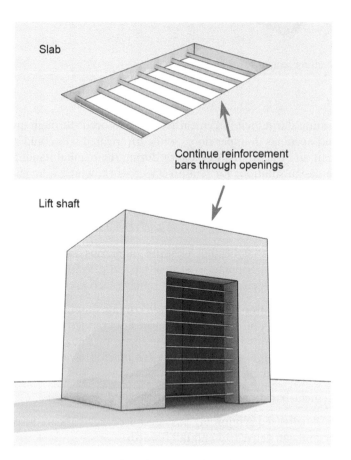

Figure 3.4 Continued reinforcement through openings.

Concrete frames

Accident statistics demonstrate that serious injuries and deaths are suffered by construction workers due to falls during concrete frame construction. Huang and Hinze (2003) calculated that 27% of falls in the construction sector in the US were attributable to concrete frame construction. The following design solutions were acquired from Thorpe (2005), WorkSafe Victoria (2008), Ghule (2008), CII (2010), OSHA Alliance Program Construction Roundtable (2010[?]a), and Speegle (2011), and validated through industry reviews, as suitable measures to reduce the risk of falls during concrete frame construction:

- Whenever craning is feasible on site:

  o specify precast members of standard sizes for upper floor slabs, beams and columns; this will eliminate workers' exposure to the risk of fall during formwork erection, reinforcement fixing and concreting;

  o when using precast panels for slab construction, design for railings to be attached to the edge panels on the ground before they are raised to a height; this will offset works on upper floor edges for fixing railings and thereby fall risks.

- If in-situ concreting is unavoidable for upper floor slabs, consider the use of permanent metal decks with concrete fill (composite decks), rather than traditional concrete slabs, to eliminate formwork and minimise rebar in elevated slabs, which present numerous fall hazards.
- Whenever windows are fixed on upper floors, consider permanent, built-in anchor points in the structural members around windows to provide firm connections for fall-arrest systems and scaffolds.

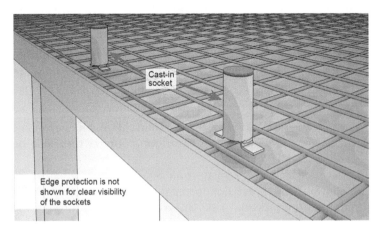

Figure 3.5 Cast-in sockets along slab edges.

- In traditional in-situ concrete slabs, specify cast-in sockets along perimeter edges of upper floors, as shown in Figure 3.5, to enable early installation of permanent railings as part of the floor, which can eliminate the need for temporary guardrails and be a safer option for work on upper floors, following slab casting.
- In designing railings or parapet walls for balconies and terraces, specify their heights to be greater than 1.10 m from the finished floor level, which eliminates the need for guardrails during maintenance work, as well as improving fall safety for building occupants, particularly little children.
- Cantilevered beams, floors and roofs are commonly used in building projects, but they can cause fall risks, depending on the height and 'extend' of their projection:

  o consider how the cantilevered sections in your design will be accessed for maintenance; would they multiply fall hazards at heights?
  o if potential risks outweigh aesthetic benefits, improve the safety of such design features: by optimising the length and height of projected parts; with built-in, permanent anchor points for safety harnesses; and with safe access routes and barriers.

Floor and roof openings

Many injuries and deaths are caused due to workers falling either from a roof edge, or through an opening in a roof, or while accessing a roof. The below-described safe design solutions to protect workers from such falls during construction, maintenance and demolition of buildings were elicited from WorkCover NSW (2009), CII (2010), OSHA Alliance Program Construction Roundtable (2010[?]d), and Safe Work Australia (2012), and verified through industry reviews.

1   Openings on upper floor slabs and roofs create greater fall risks during both construction and maintenance work. However, such openings are essential in buildings for various functional purposes, for example, stairwells, courtyards, skylights, etc.

   - in order to reduce the risks, consider grouping many small openings to form one/few large opening(s) on floors/roofs as these can be easily guarded;
   - locate floor/roof openings away from building edges as much as possible; positioning floor/roof openings near the edge of the building multiplies the potential risks as workers can fall through the openings as well as through the edge of the building during construction and maintenance work.

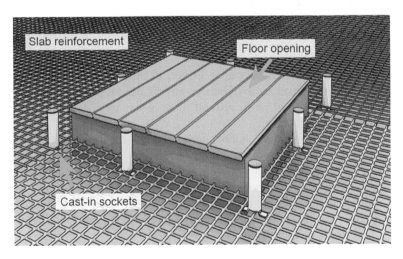

Figure 3.6 Cast-in sockets around an opening.

2   Design-in permanent cast-in sockets around floor openings, as shown in Figure 3.6, to enable early installation of permanent railings, which can provide fall protection during construction.

3   For openings that provide access to the rooftop from a level below, like roof hatches, design permanent guardrails to all sides with an auto-closing swing gate and access grab bars at the opening (see Figure 3.7).

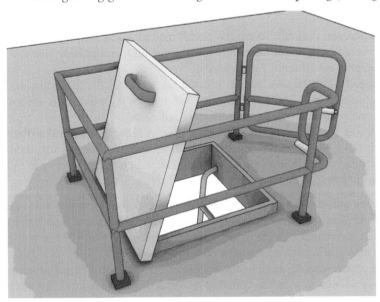

Figure 3.7 Auto-closing swing gate for openings.

4   Humps, depressions and raised levels on floors can be trip hazards, and the level of risk they present to construction and maintenance workers is increased when they are located near edges and openings. Hence:

- make covers of sumps, outlet boxes, drains, etc. level with the finished floor level to avoid these covers becoming trip hazards; introduce transition strips around these covers to make their presence easily noticeable;
- remove or permanently barricade humps, depressions, etc. around roof/floor openings, accesses and edges;
- in floor areas where a depression, raised level or hump is to remain, include some permanent warning sign, colour transitions, or another suitable design feature to make it clearly visible;
- provide adequate lighting in areas where there are depressions, raised levels or humps.

Staircases

Staircases have had a significant involvement in fall accidents not only during construction but also during occupancy of buildings. Twelve safe design suggestions for staircases are provided here, which were elicited from Sinnott (1985), Ghule (2008), CII (2010), and Australian Building Codes Board (2008), and then validated through industry reviews.

1   For staircases:

- wherever possible, consider prefabricated staircases that can be erected as a single assembly on the ground, minimising work at heights and thereby fall hazards;
- specify balustrades to be erected as part of the prefabricated staircases to eliminate the need for temporary guardrails during construction and hence eliminating fall risks;
- design and plan the erection of permanent staircases to take place in the early stage of construction to enable workers to use them instead of temporary stairs and ladders, which figure heavily in fall accidents.

2   If in-situ staircases are unavoidable, consider designing stairs that can be built using Stairmasters (see Figure 3.8), which are lightweight moulds (permanent formworks) manufactured off-site and left in place. They also come with permanent rails pre-attached.

3   In the conventional design of staircases, consider specifying staircases to have sockets cast in the concrete or welded-in steel sections on open sides/edges to enable the fixing of guardrails during construction.

Figure 3.8 Stairmaster.

(Image courtesy of Stair Master Ltd.)

4 Adopt the following safe dimensioning for staircases:

- staircases should be not less than 1000 mm wide measured between the inside edges of the handrails;
- in order to avoid people colliding with the soffit of staircases, close-out underneath areas of stairways that have headroom of less than 2100 mm high;
- a stair flight should have no fewer than 2 risers, but no more than 16 risers without a landing area;
- staircase landing length should be greater than 800 mm in the direction of the flight to provide safe rest points for users;
- if the staircase is of open-riser construction, the riser opening should be less than 100 mm to prevent children's legs from slipping through the gap.

5 Safe pitches for staircases:

- steep flights as well as low-pitched flights can cause fall accidents; a pitch of 30° is preferred as this allows the design of steps with risers and goings of adequate sizes for human foot stability when stepped on; this pitch can be achieved with a combination of a riser of 168 mm and a going of 290 mm;
- uniformity in the risers and goings of steps is necessary because the stairway user quickly gets into a rhythm of foot movements; even a small variation in the rise of steps can cause a user to stumble; similarly, small irregularities in the going can also cause the foot to be misplaced.

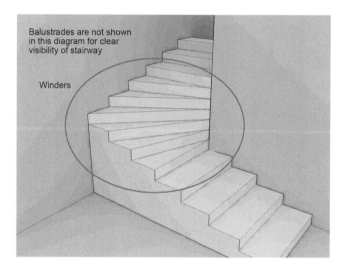

Figure 3.9 Stair winders.

6 Winders in a stairway are undesirable because they are not of adequate width at the newel end for the proper placement of the foot, as illustrated in Figure 3.9, which makes winders conducive to slips, trips and falls, especially for elderly people and for users carrying heavy or bulky objects by hand.

7 Where a staircase runs immediately in front of a doorway, a landing of adequate size should be provided near the doorway and before the start of the staircase. A landing ≥ 800 mm long between the door and the first riser is required.

8 All treads and top nosing need to have a slip-resistant finish or a non-slip strip system near the edge of each tread nosing to prevent slips and falls.

9 Lighting stairways:

- make sure that adequate daylighting is present in stairway areas through appropriate positioning of nearby windows and doors; if natural lighting is unsatisfactory in a section of a stairway, provide adequate artificial lighting so that a user will be able to see each nosing while descending or ascending the stairway;
- in order to reduce risks caused by glare that impairs the ability to see (disability glare) on stairways, tinted glass, splayed reveals or baffles may be designed-in for windows.

10 Decorations around stairways:

- lofty ceilings and decorations above stairways increase potential fall risks during installation and maintenance of such ceilings and decorations because placing a scaffold or ladder for access is difficult and hazardous; if essential, decorations should be positioned such that they are accessible from outside the stairway by other means, such as a cherry picker or scissor lift, provided that there is an access path in the building for such special equipment to reach the stairway area;

- the presence of attention-distracters, such as wall and ceiling decorations/artwork and displays around stairways, tends to cause users' eyes to automatically turn towards them, leading to loss of concentration on treads and on critical parts of a stairway, such as turns.

11 Spiral and helical staircases are aesthetically appealing but considered less safe for users relative to other staircase types. In spiral and helical staircases, the treads are at their least uniform and too narrow at the centre of the spiral or the centre void. While the non-uniform treads make the chances of slips and trips high, a user who slips on a spiral staircase may fall heavily due to its high pitch. In order to make these stairs safe, the tread/going at both ends of a step should be not less than 290 mm to enable safe stepping.

12 In designing external staircases:

- placing external staircases parallel and immediately adjacent to the building, instead of perpendicular, will minimise weather effects and thereby slips and falls;
- if positioning the staircase parallel and adjacent to the building is not possible, design an adequate covering to minimise slips and falls caused by weather effects;
- positioning external staircases on the sheltered side of the building, or under a covering, overhang or extended roofline, will protect them from the weather (rain, snow, fog, dew, etc.) and reduce fall hazards;
- siting exterior stairways on the sunny side of the building will help prevent the growth of slippery moss, which causes slips and falls;
- to guard against slipping on external staircases, treads should have a perforated or well-drained surface; moreover, finishing materials of high friction (coefficient of friction > 0.75) should be used.

Ramps

Slip, trip and fall accidents to users of ramps, including wheelchair users, can occur when the design of the ramp itself presents hazards. This section describes safe design methods for ramps, which were derived from the work of Sinnott (1985), Ghule (2008), and CII (2010), and then validated through industry reviews.

1 Adopt the following safe dimensions and material choices for ramps:

- a ramp should be at least a stride long and at least 1 m wide, have a maximum slope of 1 in 12 (approximately 5°) and a maximum length of 10 m; it should be provided with kerbs and handrails;

- particular attention is required in the choice of floor finishing for sloping surfaces; specify non-slip surface materials and finishes for ramps;
- introduce non-slip strips on ramps at appropriate intervals to prevent slips and falls.

2   Sudden, unexpected changes to the slope of a ramp may cause the failure of users to see the edges formed by the levels and sloping surfaces, resulting in trips and falls. When the slope of the ramp changes, make the edges as distinct as possible by introducing different textures or transition strips, as illustrated in Figure 3.10, to distinguish changing levels.

3   In designing external ramps:

- positioning external ramps parallel and adjacent to the building, instead of perpendicular, will minimise weather effects and thereby falls;
- if positioning the ramp parallel and adjacent to the building is not possible, minimise the possibility of slips and falls by introducing additional design features, such as roof and side coverings to prevent rainwater splash, snow, fog, dew, etc. from wetting the ramp surface;
- consider placing external ramps on the sunny side of the building to prevent the build-up of slippery moss, which causes slips and falls;
- providing a covering for external ramps by extending the roof line or by positioning them on the sheltered side of the building will reduce fall hazards caused by the weather (rain, snow, fog, dew, etc.);
- specify for external ramps a roughened or grooved surface with wide slip-resistant joints or another finish of at least equal traction; the grooves should be at an angle to the horizontal so that water does not lodge in them.

Change in ramp slope

Figure 3.10 Change in ramp slope.

Passageways and corridors

Slip, trip and fall accidents on horizontal passages and walkways in buildings could be minimised with the following safe design suggestions, as outlined in Sinnott (1985), Chubb Group of Insurance Companies (2006), Ghule (2008), and CII (2010), and validated through industry reviews.

1   Blind corners along corridors within heavy human-traffic buildings, such as commercial buildings, hospitals, institutional buildings, etc., can cause collisions and trip accidents. Introduce suitable design features, such as mirrors on the corner, to allow clear visibility of moving human traffic in all directions of the corridor.
2   Gentle slopes are safer than low-pitched steps for connecting different levels in walkways, thus ramps are to be preferred. A flight of low-pitched steps on walkways and passages requires additional attention from users to avoid slip, trip and fall accidents, but people are prone to failing to observe steps when the pitch is small.
3   Glare along walkways, passages and corridors can cause trip hazards. Thus, position windows in such a way that glare is reduced. Where there is no alternative to specific positioning of windows, the glare can be reduced by splayed reveals, tinted glass or baffles.
4   In designing external passageways:

   •   position external passageways parallel and adjacent, instead of perpendicular, to the building for minimising weather effects and thereby fall accidents;
   •   consider placing external passageways on the sunny side of the building to prevent the build-up of slippery moss, which causes slips and falls;
   •   provide a covering for external passageways by extending the roof line or by positioning them on the sheltered side of the building to reduce fall hazards caused by the weather (rain, snow, fog, dew, etc.);
   •   prevent ponding, and thereby fall hazards, on external passageways with a small gradient.

Railings

The below-listed safe design methods are suggested for designing railings that are attached to staircases, ramps and walkways. These suggestions were elicited from Sinnott (1985), Ghule (2008), CII (2010), and Australian Building Codes Board (2008), and refined through industry reviews.

1   Adopt the following design considerations for safe balustrades:

   •   a balustrade is required along all open sides and edges of any staircase, balcony, ramp, hallway, veranda, mezzanine floor or bridge where the potential to fall from the above-mentioned areas is 1000 mm or greater to the surface below;

- the height of the balustrade on a finished floor, landing, balcony or on an elevated path should be greater than 1100 mm; a balustrade of this height will eliminate the need for a guardrail during construction and maintenance, thereby enhancing safety throughout;
- designs should minimise the risk of children falling or slipping through gaps in balustrades; the widest gap in balustrades should not exceed 100 mm;
- horizontal members in balustrades that can facilitate children climbing should be avoided;
- design handrails and top rails to support a minimum load of 90 kg applied within 50 mm in any direction (downward/outward) and at any point along the railing;
- design intermediate vertical members of balustrades to have a maximum spacing of 475 mm;
- maintain a consistent railing height throughout the project.

2   Provide railings for steps and walkways that have no fewer than four risers, or a rise of 750mm in height (whichever is less). Provide railings for landings even when they are only a few risers up (see Figure 3.11).

3   For safe handrails:

- use a tubular rail with a diameter of 45–50 mm, or a shaped rail of 65 mm width, so that anyone tripping on the staircase or ramp can grip the handrail comfortably (see Figure 3.12);
- for gripping in an emergency, the handrail needs to have a clearance of at least 65 mm from the face of the wall;
- on wooden stair rails, fix the top rail on top of the vertical members or on the inside of the members, rather than on the outside of them;
- avoid any lever-shaped ends for the stair rail and other objects/ fixtures on the top rail as they may catch the clothes of climbers and cause trip hazards; make joints and railing ends round and even.

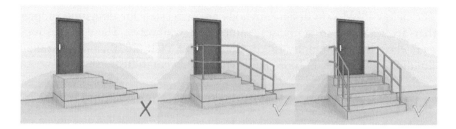

Figure 3.11 Low-rise platforms.

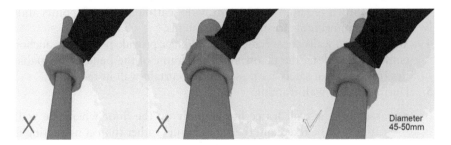

Figure 3.12 Safe handrail size.

Walls

Huang and Hinze (2003) reported that approximately 5% of fall accidents in construction are attributable to masonry works, while Dong et al. (2009) discovered that about 4% of construction fatalities are suffered by masonry contractors. An industry validation of the findings from a detailed content analysis of Culvenor (2006), Ghule (2008), Safe Work Australia (2012), and CII (2010) recommends the following safe design solutions for reducing fall accidents during masonry works:

1   Consider precast concrete, lightweight panels or modular components for walls, particularly for external walls on upper floors, rather than traditional masonry. These will reduce work on upper floor edges and thereby fall risks.
2   If traditional masonry walling is involved, specify lightweight, consistently shaped and sized masonry bricks/blocks to make manual handling at heights safe.

Figure 3.13 Anchor bolts on a wall.

3   Design-in permanent anchor bolts or holes into external walls of buildings, as shown in Figure 3.13, to provide scaffolding tie-off points and/or stable connections for fall-arrest systems.
4   When curtain walls are specified for buildings, provide permanent anchor points or ties along the perimeter of the frame of the building to enable the attachment of safety harnesses during curtain-wall installations.
5   During timber wall framing:

- assemble the wall frame horizontally on the floor where the wall frame is to be fixed and then stand it up, rather than constructing it vertically at heights with individual timber members;
- prior to standing up wall frames for external walls, fix guardrails in window and door openings up to 1.10 m high from the floor level.

Doors

Openings in walls for external doors and inappropriately placed internal doors present fall hazards for construction workers and building occupants. In order to eliminate such fall hazards, the below-presented safe design suggestions were developed by analysing Sinnott (1985), Ghule (2008), CII (2010), and Safe Work Australia (2011), and then validated through industry reviews.

1   Provide protection/overhangs/canopies of adequate sizes above external doors to eliminate slippery conditions caused by rain, snow, dew, fog, etc.
2   Consider providing inserts in exterior door jambs on upper floors for installing guardrails during construction and maintenance (see Figure 3.14).

Figure 3.14 Inserts in door jambs.

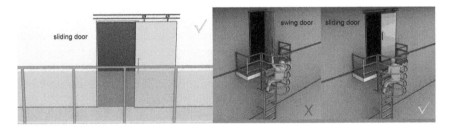

Figure 3.15 Sliding doors for overhead rooms and corridors.

3 Where practical, include a vision panel of 1000 mm high in swing doors, starting from 750 mm above the floor level, to allow forward vision of what lies on the other side for anyone approaching the door.

4 Design sliding doors along corridors, passageways and balconies, rather than swing doors (see Figure 3.15). Swing doors in these spaces can cause struck-by and fall accidents to users.

5 Eliminate trip hazards such as raised door sills and immediate level changes on the floor around doors, except where a level change is normally anticipated, such as at the external face of a front door in a building.

6 To avoid trips and falls caused by clothes being caught by door handles, use door handles that return to within about 3 mm of the face of the door, or handles that do not have a lever.

7 Keep doorways away from the bottom of a stair flight by leaving a minimum space of 800 mm long between the door swing and the stair flight (see Figure 3.16).

8 Consider locating doorways away from stairway heads as doorways directly opposite to a stairway head present fall hazards during construction, as well as during occupancy, of a building.

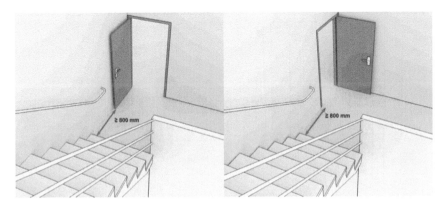

Figure 3.16 Doors near stairways.

Windows

Openings for windows in external walls of upper floors present fall hazards for construction and maintenance workers as well as for building occupants. Safe design suggestions are presented in this section to reduce the incidence of falls through window openings, which were derived from a content analysis of Sinnott (1985), Ghule (2008), CII (2010), and Safe Work Australia (2011), and then refined through industry reviews.

1 Design windows that can be installed and maintained from inside the building. Moreover, make the members of the windows and the glazing systems strong enough to withstand accidental and casual loading equivalent to that which would occur if a person fell against a member. This will help to reduce the possibility of falls caused by failing window members when forces are applied on them by construction or maintenance workers.
2 Consider designing inserts in window jambs on upper floors for fitting guardrails during the installation and maintenance of windows (see Figure 3.17).
3 Add anchor bolts along the building perimeter to provide lifeline tie-off points for window installation and maintenance work.
4 Falls of children through openable windows in occupied, multi-storeyed residential buildings have been reported. To make openable windows as safe as possible, consider the following design options:

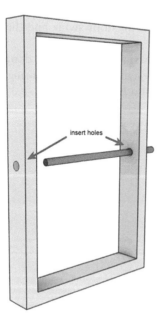

Figure 3.17 Inserts in window jambs.

- install permanent grilles on the inside face of the window, which could be opened in emergency situations;
- make the height to the bottom sill of the openable window well above that which a young child can reach by standing on a chair, that is, design window sills of 1.10 m high from the floor level; window sills of this height will function as guardrails during construction and maintenance work, and will improve child safety during occupancy;
- make sure that an inner sill or window ledge does not provide a platform for a child to stand on;
- if permanent grilles and high sills are not allowed, design-in self-contained mini-alarms to give warning whenever the window is opened, especially by children.

Floor finishes

The type of floor finish used within different parts of a building is a vital consideration for minimising slip and fall accidents during occupancy and maintenance. Important factors to be considered in the selection of floor finishes are provided below, based on Chubb Group of Insurance Companies (2006), CII (2010), and Speegle (2011), as well as the views of industry experts.

1 Slip and fall accidents occur on floors due to the level of friction between the heel of the user and the floor. The safest Coefficient of Friction (COF) is 0.22 for men, and 0.19 for women, in normal straight walking. However, the contribution of the person to the COF value is largely dependent on the footwear. The contribution of the floor surface is affected by wear, wet conditions and maintenance. Hence, it is advisable to choose floor finishes that have COF values of higher than 0.50 for internal floor surfaces.

2 As different floor finishes have different COF values, when multiple finishing materials are used for one floor, users need to be aware of them to negotiate a safe rhythm of walking across differently finished floors – likewise, for small level variations on a floor with the same finish. Clearly differentiate variations in floor finishing or levels with transition strips to make the changes easily visible to building users.

3 For external floor surfaces and for locations where wet conditions are likely, choose floor finishes that have COF values of higher than 0.75, for example:

- specify matt or light broom-finish concrete for surfaces that are exposed to weather;
- introduce safe walking surface joint covers at floor expansion joints.

4 Be aware of trip and fall hazards on split-level floors. The design of split-level floors in residential buildings is common and home occupants

Split floors with different finishes
and transition strip on common
vertical face

Figure 3.18 Split floor.

are aware of their surroundings. However, in other types of building, such as hospitals, schools, elderly/child care centres, etc., where human traffic is heavy and the user mix varies considerably, split-level floors can cause trip and fall accidents. Avoiding split levels in such building types will eliminate such risks. If split-level floors, however, are needed, consider specifying different floor finishes/colours for split levels and/or transition strips at their intersections to make the level difference clearly apparent to building users (see Figure 3.18).

Pitched roofs

Falling from roofs constitutes the most common fall accident as roof works are invariably conducted at heights and they normally involve more potential hazards, such as sloped surfaces, unguarded roof openings and edges, and fragile roof materials (Suruda, Fosbroke and Braddee 1995; Holt 2005, p. 228; Huang and Hinze 2003). The below-described safe design methods for pitched roofs were collated from Safe Work Australia (2011), Ghule (2008), WorkCover NSW (2009), OSHA Alliance Program Construction Roundtable (2010[?]a, c), CII (2010), and Safe Work Australia (2012), and then validated through industry reviews.

1 High-pitch roofs can cause severe slip and fall accidents to construction, maintenance and demolition workers. Consider designing low-pitch roofs to reduce such risks (design roofs with a pitch of less than 30°).
2 For roof structures:

  • where practical, specify prefabricated/pre-assembled structural elements (e.g. trusses), rather than individual members, which can be placed by crane with minimal work at heights;

- design roof structures with a maximum spacing of 600 mm between trusses, rafters and battens to prevent workers from falling through the gaps.

3   Where practical, consider including a permanent safety mesh underneath the roof, which will halt the falling of workers to the ground/level below.

4   Where a part of the roof is covered with fragile materials, such as fibreglass panels, a permanent warning sign should be fixed near the area. Moreover, the fragile roof area should be enclosed by permanent guardrails.

5   To make roof-maintenance work safe:

- consider specifying roofing materials, including seals, that do not require regular inspection, maintenance and replacement;
- install permanent anchorage points or anchor bolts on pitched roofs (see Figure 3.19) in order to provide permanent tie-off points for fall-arrest systems during construction and maintenance work;
- design-in permanent location brackets or ladder-tying points along the roof edge to prevent ladders from slipping sideways during roof construction and maintenance (see Figure 3.19).

6   For roofs with rooftop equipment:

- where workers need to have access to a roof for maintenance work and mechanical equipment on the rooftop, permanent stairways, or fixed ladders with a safety cage, and walkways should be provided;
- fixed ladders are usually attached upright, but ladders at 15° to the face of the building are ergonomically better for climbing and safety in use;
- if a walkway on a roof is to be designed-in for roof or rooftop-equipment maintenance, the requirements of the relevant construction standard for rooftop walkways must be followed; the walkways must have the following minimum requirements: width > 450 mm; guardrails ≥ 1.10 m; there must be a toe board of at least 160 mm high; maximum spacing between the toe board and an intermediate rail should be < 0.75 m.

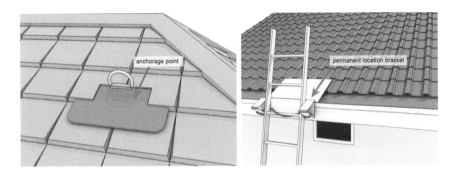

Figure 3.19 Roof anchorage point and ladder location bracket.

Source: left image modified from OSHA Alliance Program Construction Roundtable 2010[?]c.

Concrete flat roofs

Safe design solutions for concrete flat roofs from OSHA Alliance Program Construction Roundtable (2010[?]a, c), CII (2010), and Safe Work Australia (2012), which were also validated through industry reviews, include the following:

- Concrete flat roofs present serious fall risks for maintenance workers. In order to reduce these risks (see Figure 3.20):
  - o   design parapet walls or railings for roofs to be at least 1.10 m high so that they can function as permanent guardrails during the construction and maintenance of roofs;
  - o   avoid designing elevated exterior structures, equipment, etc. next to roof edges;
  - o   install permanent anchorage points on roof slabs to provide permanent tie-off points for fall-arrest systems during construction and maintenance work.

- Often, HVAC equipment and satellite dishes are placed on the rooftop in multi-storeyed buildings. Maintenance workers are faced with numerous fall hazards because of their location:
  - o   where practical, place HVAC equipment and satellite dishes on the ground;
  - o   if rooftop equipment is unavoidable, locate it away from roof edges, openings and skylights.

Complex roofs

For complex-shaped roofs, as in Figure 3.21, it is harder and riskier to carry out formwork at heights, and then work on the complex-shaped formwork to fix reinforcement, and then cast concrete, than to perform those tasks on

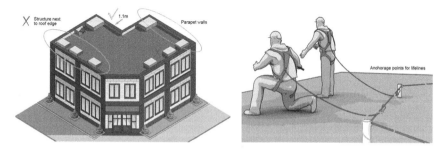

Figure 3.20 Flat roof safety measures.

(Left image copyright purchased from iStock and adapted.)

flat roofs. Moreover, climbing on these complex roofs for maintenance work is also riskier. The movement of a cherry picker boom around the roof, and thereby access to different parts of the roof, may be limited due to the multiplicity and complex connectivity of roof units. The below-described safe design methods may be adopted when a complex roof is an essential design feature in a project.

1 Consider the use of precast elements to build complex roofs as these will eliminate the need for formwork, reinforcement and concreting at heights. The magnitude of work at heights is significantly reduced as the elements are cast on the ground and fixed in place by craning with minimum work at heights.
2 Design-in an adequate number of anchor points on the precast roof units to provide connection points for fall-arrest systems during maintenance.
3 Introduce permanent walkways and access routes on the precast roof units to facilitate workers' safe access and movements during maintenance.

Figure 3.21 Complex roof.

(Image copyright purchased from iStock and adapted.)

Skylights

Skylights are installed on roofs as a green design and energy-conservation measure. However, deaths or serious injuries due to workers falling through skylights have been repeatedly reported in OHS statistics. In the US, for instance, annual fatalities during skylight work ranged from 18 to 36 between 2003 and 2006 (BLS 2008 as cited in OSHA Alliance Program Construction Roundtable 2010e). Design solutions to protect against the risk of falls through skylights during construction, maintenance and demo-lition of roofs were elicited from Ghule (2008), WorkCover NSW (2009), the OSHA Alliance Program Construction Roundtable (2010e), and CII (2010), and are described below. These suggestions were also reviewed and refined by industry experts.

1   In designing skylights for roofs:

- position skylights on a raised curb of 250–300 mm high so as to clearly indicate the presence of an opening and to avoid workers accidentally walking into it (see Figure 3.22);
- design domed/pyramidal, instead of flat, skylights with non-breakable glass; domed or pyramidal skylights are less likely to be accidentally stepped on;
- specify skylight types that are capable of withstanding the live load of a worker who accidentally steps or falls on the skylight.

2   Design permanent guardrails around flat skylights, or skylights with low convexity, to prevent construction and maintenance workers from accidentally stepping or falling on them.

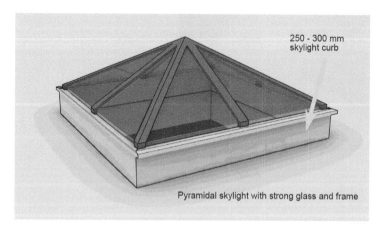

Figure 3.22 Skylight placement.

Source: modified from OSHA Alliance Program Construction Roundtable 2010e.

3   Place skylights away from:

- roof edges to prevent workers from falling through roof edges during skylight installation and maintenance;
- rooftop mechanical/HVAC/communication equipment to prevent
  the likelihood of workers falling on/through skylights.

Roof vents and gutters

Fall accidents may happen during the installation and maintenance of roof
vents and gutter systems. This section has design suggestions to reduce
such risks.

1   For roof vents and exhausts:

- combine small exhaust vents into a single, large vent;
- mount roof vents at a low level or route them through the sides of
  the building, instead of being mounted on the roof.

2   For roof gutter systems:

- specify large-volume gutters and downpipes to reduce the number of gutters and downpipes installed; this will minimise work at
  heights for their installation and maintenance;
- introduce top covers for gutters, which will reduce the amount of
  regular cleaning, and thereby work at heights or on the roof;
- position gutters so that their installation and regular maintenance
  can be performed safely from safe access routes or from mechanical
  equipment such as cherry pickers.

Ceilings

Dong et al. (2009), and Huang and Hinze (2003) recorded serious falls
from heights during the installation of ceilings. Safe design suggestions for
ceilings from Speegle (2011), and CII (2010), as refined through industry
reviews, include the following:

- In designing suspended ceilings:
  - consider simplifying the design as complex ceiling systems make
    installation and maintenance work at heights more hazardous;
  - specify ceiling hangers and connections that are capable of support-
    ing construction live loads, including the weight of a worker;
  - specify smaller, lightweight ceiling materials for safer handling
    during installation and maintenance at heights;
  - specify and schedule ceiling materials to be painted and/or insu-
    lated on the ground prior to their placement to reduce ceiling
    works at heights.

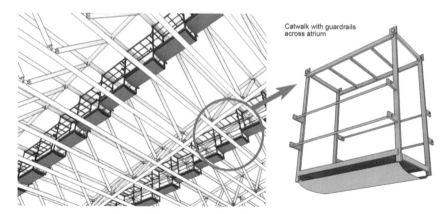

Figure 3.23 Catwalk across an atrium.

- When designing an atrium or a high ceiling for a building:
  - consider including a permanent, safe gantry, or catwalk of adequate size, with guardrails across the atrium to provide a safe working platform (see Figure 3.23);
  - install permanent anchor points that can be used by workers during ceiling work.

Skyrise greenery

Skyrise greenery is embraced in urban developments as a sustainable design feature that integrates greenery into high-density urban spaces. It is implemented as one of two forms: rooftop greenery, and/or vertical greenery. Rooftop greenery refers to horizontal planting work on roof slabs, whereas vertical greenery is formed by greening facades and ledges of buildings by installing vegetation. Figure 3.24 illustrates the presence of rooftop greenery on a building.

While skyrise greenery could bring many environmental and aesthetic benefits to buildings, its installation and continual maintenance present numerous fall hazards due to the precarious locations for such work. Behm and Hock (2012) proposed the following safe design solutions for skyrise greenery systems to reduce risks for workers.

1 The load-bearing capacity computations for roofs with greenery must consider the potential maximum load of greenery, and also the load of soil and equipment used for installing and maintaining greenery, in addition to the usual dead and live loads.
2 Permanently fixed stairs of adequate size and strength should be provided through the building's core for workers' safe access to rooftop greenery areas for installation and maintenance of plants.

Figure 3.24 Skyrise greenery.

(Image copyright purchased from iStock and adapted.)

3   Permanent parapet walls or railings of greater than 1.10 m height should be provided around rooftop edges.
4   Design-in permanent restrain lines, housed in boxes to protect from weathering, to provide connection points for safety harnesses to enable workers' free and safe manoeuvre during installation and maintenance of greenery.
5   Plants that hang over the edges of roofs or ledges pose significant fall hazards during maintenance. Therefore, select plant types for roofs and ledges that are not creeper plants or wider shrubs.
6   Design and locate vertical greenery systems (green walls and ledges) such that they can be accessed from the ground, or from a safe platform, for maintenance. Alternatively, innovative design solutions such as movable/rotating greenery systems, hinged greenery modules and rear-access greenery systems could be considered.
7   When ledge greenery is designed, consider fall protection and safe access for installation and maintenance of plants, that is, design-in horizontal restrain lines behind the vegetation; also allow sufficient space between the back wall and plants, and between individual plants.
8   Consider maintenance activities such as weeding, watering, fertilisation, pruning, replacement, disease control, etc. and select plant types that do not require very frequent maintenance.

Steel structures

The erection of steel structures often involves work at heights on incomplete structures and often requires the use of machinery. About 36 fatalities per year have been reported in the US as being caused by falls from structural steel during the construction of buildings (BLS 2008). These direct falls of steel erectors occur due to inadequate fall-protection systems, difficulties in handling loads, and loss of balance at heights. Moreover, another set of accident records related to steel construction suggests that falls leading to deaths and serious injuries of steel erectors occurred due to the failure of the structural elements that they were working on. Building codes require design and sizing of structural members based on their dead load and the live load resulting from building occupancy. However, the failure of structural elements has been known to occur because the additional loads caused by construction machinery, materials and processes have not been considered by designers (OSHA Alliance Program Construction Roundtable 2010[?]g).

A content analysis of Thorpe (2005), Ghule (2008), WorkCover NSW (2009), OSHA Alliance Program Construction Roundtable (2010[?]f, g), and CII (2010) and a subsequent industry-review process resulted in several design solutions, as described below, that can reduce the risk of serious falls from structural steel during construction.

1 Specify maximum prefabrication of structural members/systems to minimise the amount of overhead work and erection duration to reduce fall hazards.
2 When steel-framed structures are not laterally braced, restrict their heights such that the stability of the structure during erection is assured.
3 Specify steel-framing systems that reduce the number of individual structural elements.
4 Design steel-framing systems with beams that are fully rested on supports, rather than being attached to the sides of columns or other members by cleats, plates and/or bolts.
5 When sizing structural members, consider the temporary load resulting from the weight of construction machinery, materials and construction forces involved in the erection of the structure, in addition to the dead load of members and the live load from building occupancy.
6 Provide information on load-bearing capacities of structural members, indicating how much additional load can be tolerated should temporary equipment and other loads be placed on them.
7 Specify structural members of consistent size, preferably lightweight and easy to handle.
8 Size the depths of structural members so that there is sufficient headroom clearance around staircases, platforms, and all areas of access and exit.
9 Specify a minimum beam width of 150 mm to allow a sufficient walking surface for steel erectors.
10 Design floor plates/decks to be placed as early as possible in the construction period so that the reliance on fall-arrest systems, working platforms and ladders can be minimised.

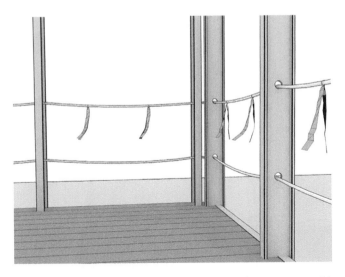

Figure 3.25 Pre-drilled holes on steel columns for perimeter cables.

Source: adapted from OSHA Alliance Program Construction Roundtable 2010[?]f.

11 Ensure that access staircases form part of the early frame in order to reduce the reliance on ladders and beams for access.

12 Specify holes be drilled in perimeter columns during prefabrication at appropriate heights (preferably at 0.55 m and 1.10 m) above the floor level, as shown in Figure 3.25, to enable easy installation of perimeter safety cables as soon as floor decking has been installed.

13 Similarly, specify holes be drilled during prefabrication at appropriate intervals in perimeter beams and columns, and in beams around floor openings, for use by steel erectors as anchor points for fall-arrest systems.

14 Specify holes be drilled during prefabrication in the webs of beams placed over service piping and duct works to enable the attachment of pipes, ducts and fall-arrest systems.

15 Clearly annotate construction drawings to provide details of the location of holes for anchor points and perimeter safety cables along beams and columns.

16 Specify pre-attached cleats on columns at column–beam connections (see Figure 3.26), rather than welding on site, so that the ends of beams can rest during the connection process and thus make it safer and easier for the steel erectors.

17 Minimise cantilevered beams and floors. If cantilevered floors and beams are designed for functional requirements, optimise their projection from the building edge so that the risks of falls from edges are not amplified during construction, maintenance and/or later demolition of such parts.

Figure 3.26 Pre-attached cleat.

Source: adapted from OSHA Alliance Program Construction Roundtable 2010[?]f.

18 Specify column splice connections to occur at two-floor intervals, and position splices at about 1.2 m above the floor level.
19 Specify a single size, or only a few sizes, for bolts, nails and screws. If several sizes are needed, specify sizes that can be easily differentiated.
20 Make sure that patterns of bolt holes are uniform throughout the frame for easy insertion of bolts.
21 Specify a minimum of two bolts, nails or screws per connection, and provide an extra dummy hole at bolted beam–column connections for inserting a spud wrench or other objects to provide support during installation.

Conclusion

Two primary challenges for the successful adoption of PtD in the industry are: the still-developing body of PtD knowledge, and the limited PtD competencies of designers. These two are interrelated issues in that a developed body of PtD knowledge is likely to uplift the competency level of designers in PtD. In a bid to address this challenge, this chapter has established contents for a well-structured knowledge base of safe design suggestions for fall prevention in construction. The new knowledge base can lead to two significant practical implications: first, it can enable the development of web-based systems or mobile computing tools to enhance designers' PtD competency and thereby support faster adoption of PtD in the industry; second, it can provide the basic ingredients and rules necessary for creating expert systems or virtual reality/BIM applications that are capable of auditing building designs for accident Prevention through Design features. The next chapter focuses on utilising the new knowledge base for developing a mobile computing tool for fall PtD.

References

Australian Building Codes Board. (2008) *Building Code of Australia, Volume 2.* Canberra: Australian Building Codes Board.

Behm M and Hock P C. (2012) Safe design of skyrise greenery in Singapore. *Smart and Sustainable Built Environment*, 1(2): 186–205.

Chubb Group of Insurance Companies. (2006) *Preventing Slip-and-Fall Accidents.* URL (accessed 12 Mar. 2014): http://www.chubb.com/businesses/csi/chubb4992.pdf.

Construction Industry Institute (CII). (2010) *Design for Construction Safety Toolbox, Version 2.0.* Texas: Construction Industry Institute.

Culvenor J F. (2006) *Creating an environment for transformational change: The role of culture and innovation in risk management practice.* Keynote presentation at Risk Management Research and Practice: An Educational Perspective, University of Wales, Holyhead, Anglesey, UK, 30–31 March 2006.

Dong X S, Fujimoto A, Ringen K and Men Y. (2009) Fatal falls among Hispanic construction workers. *Accident Analysis and Prevention*, 41(2009): 1047–1052.

Dong X S, Choi S D, Borchardt J G, Wang X and Largay J A. (2013) Fatal falls from roofs among US construction workers. *Journal of Safety Research*, 44(1): 17–24.

Ghule S. (2008) *Suggested Practices for Preventing Construction Worker Falls.* Master's Thesis, University of Florida.

Holt A. (2005) *Principles of Construction Safety.* Oxford: Blackwell Science Ltd.

Huang X and Hinze J. (2003) Analysis of construction worker fall accidents. *Journal of Construction Engineering and Management*, 129(3): 262–271.

Lim B. (2014) Vivo city Singapore. [Photograph] (Lim's own private collection).

OSHA Alliance Program Construction Roundtable. (2010[?]a) *Falls from Floor Openings.* URL (accessed 10 July 2013): http://www.designforconstructionsafety. org/media.shtml

OSHA Alliance Program Construction Roundtable. (2010[?]b) *Falls from Ladders.* URL (accessed 12 July 2013): http://www.designforconstructionsafety.org/ media.shtml

OSHA Alliance Program Construction Roundtable. (2010[?]c) *Roof Anchors.* URL (accessed 13 July 2013): http://www.designforconstructionsafety.org/media.shtml

OSHA Alliance Program Construction Roundtable. (2010[?]d) *Roof Hatch Access and Hole Protection.* URL (accessed 15 July 2013): http://www.designforconstructionsafety. org/media.shtml

OSHA Alliance Program Construction Roundtable. (2010e) *Specify Non-fragile Skylights and/or Skylight Guards.* URL (accessed 15 July 2013): http://www. designforconstructionsafety.org/media.shtml

OSHA Alliance Program Construction Roundtable. (2010[?]f) *Structural Steel: Beams and Columns.* URL (accessed 20 July 2013): http://www.designforconstructionsafety. org/media.shtml

OSHA Alliance Program Construction Roundtable. (2010[?]g) *Structural Steel: Construction Loads.* URL (accessed 20 July 2013): http://www.designforconstruction safety.org/media.shtml.

Safe Work Australia. (2011) *Managing the Risk of Falls at Workplaces.* URL (accessed 20 Aug. 2013): http://www.safeworkaustralia.gov.au/sites/SWA/about/Publications/Documents/632/Managing_the_Risk_of_Falls_at_Workplaces.pdf

Safe Work Australia. (2012) *Preventing Falls in Housing Construction: Code of Practice.* URL (accessed 6 Feb. 2013): http://www.safeworkaustralia.gov.au/sites/SWA/about/Publications/Documents/694/Preventing%20Falls%20in%20Housing%20Construction%20Read%20only.doc.

Sinnott R. (1985) *Safety and Security in Building Design.* London: Collins.

Speegle A. (2011) *1600+ PtD List.* URL (accessed 20 July 2013): http://www.designforconstructionsafety.org/media.shtml

Suruda A, Fosbroke D and Braddee R. (1995) Fatal work-related falls from roofs. *Journal of Safety Research* 26(1): 1–8.

Thorpe B. (2005) *Health and Safety in Construction Design.* Gower Publishing Limited, England.

WorkCover NSW. (2009) *Safe Design of Buildings and Structures.* URL (accessed 11 July 2013): http://www.workcover.nsw.gov.au/formspublications/publications/Documents/safe_design_buildings_structures_2088.pdf.

WorkSafe Victoria. (2008) *Prevention of Falls in General Construction.* URL (accessed 23 Feb. 2013): http://www.worksafe.vic.gov.au/__data/assets/pdf_file/0015/9231/cc_fallsconst_web.pdf.

4 Mobile app development for fall Prevention through Design

Introduction

Mobile apps have become one of the common commodities today, so much so that there are apps for almost every activity people do. They are commonly used by people of any age group or educational background. In the professional world too, mobile apps are being introduced to supplement work practices or existing information and communications technology (ICT) systems. The simplicity, cost-effectiveness and wider accessibility are drivers for the proliferation of mobile apps in the industry as well as in normal life. It is therefore regarded as integral for any business today to develop mobile apps that support them.

This chapter aims to expound the development process of a mobile app for fall Prevention through Design (PtD) in construction. The chapter starts with an account of mobile apps used in construction. Then, different mobile app development models and cycles are discussed. Next, the process involved in developing the mobile app for fall PtD is elaborated. Following that, the use of the fall PtD mobile app and the lessons learnt from its development are explained. Finally, the chapter is summarised, thus setting a context for the next chapter.

Mobile apps in construction

Construction is an information-intensive industry due to the significant number of its stakeholders (architects, engineers, builders, contractors, subcontractors, suppliers, etc.), the long time frame of projects, the complex nature of the process, and the dynamic nature of construction sites. The information generated is typically retrieved remotely from a range of locations and conditions. Current practice within the field involves information access, editing and decision making that is often limited to 2D paper-based technical drawings (Anumba and Wang 2012, p. 1). Stakeholders demand and rely on accurate and on-time communication. However, the process suffers efficiency problems contributing to significant time and cost issues, an area that mobile computing and ICT have the potential to address. The technology provides possibilities for increased speed and improved quality of communication; improved accessibility and exchange of information; as well as allowing for more effectiveiy distributed teamwork, thereby increasing efficiency and productivity, and delivering cost savings (Koseoglu and Nielsen 2005; Chen and Kamara 2008).

Mobile computing covers three broad areas: portable hardware, the communication networks, and software. Ahsan et al. (2007) discussed the main components as hardware, network, and services, where services correspond to applications and tools available on mobile devices. Portable hardware includes tablet PCs, personal digital assistants (PDAs), pocket computers, wearable computers and smartphones. Mobile phones are the key hardware tools used on construction sites for voice and SMS communications, while desktop computers or laptops are the main devices employed in offices for email (Ahsan et al. 2007, p. 259). Networks provide the vital component to mobile computing, enabling internet access to collaboration tools as well as access to email and other data. For outdoor environments, the significant workspace within construction, wireless networks tend to offer the best solution (Ahsan et al. 2007, p. 260).

Current mobile phone and tablet technology has significant power and supports a large range of applications previously restricted to desktop computers. The smartphone devices and apps are becoming standard ways of communicating and working within the field of construction. Mobile apps offer possibilities of improved workplace efficiency through access and sharing of information, as well as assisting in the logistics of construction work and management. These apps are utilised for opening and viewing of data files such as BIM or CAD models (e.g. BIMx, AutoCAD WS, TurboViewer), calculations and material estimates (e.g. DeWALT Mobile Pro, Home Builder Pro Calcs), project management and collaboration (e.g. Aconex collaboration platform), bid management (e.g. SmartBidNet) and administration of other aspects of the construction process including financial, equipment and people (e.g. Maxwell Systems ProContractorMX). A key benefit of the apps available for project management is the ability to have up-to-date drawings or other information on screen, material that was previously distributed only in a paper format. Since many different people on site, or even within the office environment, share digital versions of drawings, there is potential for significant paper reduction and increased accuracy in communication. The mobile apps for smartphones and tablets even cater for construction-tool functionality such as levels (e.g. sightLevel) or rulers (e.g. i-Ruler). Table 4.1 shows a range of mobile apps currently available to those within the construction industry. Yet, apps for managing workplace health and safety in construction is an area that has been little explored.

Development of mobile applications

The stages of design and development of mobile applications are similar to that of a dynamic website, desktop multimedia application or even a game.

Table 4.1 Mobile apps for construction

App Name	Platforms	Cost	Description
Aconex Mobile (Aconex 2013)	iPad, iPhone, iPod Touch (Android coming soon)	Free (works with paid subscription to Aconex)	Aconex is a project collaboration and online document management platform, which enables project teams to upload, share, view and review construction documents and drawings. It also allows capturing and transmitting media such as photos, videos, and audio. Site tasks, issues and actions can also be managed through the platform. It allows offline work and sync back to Aconex once connection is restored.
AutoCAD 360 (Autodesk 2013a)	Android, iPad, iPhone, iPod Touch	Free	AutoCAD 360 allows users to view, edit (with basic functionality) and share AutoCAD files. Files can be shared and accessed through AutoCAD's web service or others like Google Drive.
BIMx (Graphisoft 2013)	Android, iPad, iPhone, iPod Touch	Free	BIMx provides a 3D interactive environment to allow users to explore fully rendered 3D Building Information Model (BIM) files.
Buzzsaw (Autodesk 2013b)	Android, iPad, iPhone, iPod Touch	Free (works with a paid Buzzsaw cloud service)	Buzzsaw allows users to securely access and view 2D and 3D DWF files, documents, images, office documents and images, and upload files to the Buzzsaw cloud.
DeWALT Mobile Pro (DeWALT 2013)	iPad, iPhone, iPod Touch	Free (paid add-ons available)	DeWALT Mobile provides calculators and conversion tools, as well as reference materials. It includes templates for calculating areas, lengths, and volumes, as well as estimating quantities of studs, drywalls and concrete slabs.
FormMobi (a la mode 2013)	Android, iPad, iPhone, iPod Touch	Part of paid monthly FormMobi plan	FormMobi acts as a virtual clipboard, allowing users to gather and distribute data through custom or pre-built forms, synced to a FormMobi cloud account.
Home Builder Pro Calcs (Double Dog Studios 2013)	iPad, iPhone, iPod Touch	US$5.49	Home Builder Pro provides guides, performs cost estimates and quantity take-off. Calculation results can be saved and shared via email.
i-Ruler (Idan Sheetrit 2013)	iPad, iPhone, iPod Touch	Free (paid versions beyond 55 cm (21 in.) available)	i-Ruler provides a ruler for iOS devices.

(continued)

Table 4.1 (continued)

App Name	Platforms	Cost	Description
ProContractorMX (Maxwell Systems 2013)	iPad	Free app works with paid ProContractorMX software	ProContractorMX is a comprehensive construction management software solution offering key capabilities for managing take-off, estimating, bidding, projects, finances, procurement, inventory, employees, payroll, and equipment. It also features intelligent dashboards and critical reports. The Mobile Connect app provides full access to project documents and data.
PlanGrid (Loupe 2013)	iPad, iPhone, iPod Touch	Free for first 50 documents; monthly plans available for additional usage	PlanGrid is a cloud-based app to store and share project documents in PDF format. It allows notes to be added, updating in real time for all users.
ProntoForms (TrueContext 2013)	Android, iPad, iPhone, iPod Touch, Blackberry, HP webOS, Windows Phone	Part of paid monthly ProntoForms subscription plan	ProntoForms allows users to create mobile forms or choose from templates for entry of field data on site.
Onsite Planroom (UDA Technologies 2013)	Android, iPad, iPhone, iPod Touch	Free (works with ConstructionOnline. com – free account offers 1GB of storage space)	Onsite Planroom allows review and sharing of plans and construction documents synced through a ConstructionOnline cloud account. It supports PDF, Word, Excel, RTF and image files.
sightLevel (Auman Software 2013)	iPad, iPhone, iPod Touch	Free (optional paid 'Pro' upgrade available)	sightLevel provides the ability of iOS devices to act as a virtual laser level, allowing measurement of the tilt of a structure through the camera of the mobile device.
SmartBidNet (JBKnowledge Technologies 2013)	Android (in beta), iPad, iPhone, iPod Touch	Free for paying subscribers to SmartBidNet	SmartBidNet is a construction bid tool to simplify and automate pre-construction, including pre-qualification, subcontractor management, invitation to bid, project document collaboration and distribution functionalities.
TurboViewer (IMSI/ Design LLC 2013)	Android, iPad, iPhone, iPod Touch	Free (optional paid 'X' and 'Pro' upgrades available)	TurboViewer allows viewing of CAD DWG files in 2D and 3D, allowing zooming and rotating of 3D models.

All these product types require planning, development and testing. In *Mobile Applications: Architecture, Design and Development*, Lee, Schneider and Schell (2004) defined the five phases of activities to work through as being the Requirements Phase, Design Phase, Coding Phase, System Test Phase, and Deployment Phase. On the surface, these phases do not indicate any unique considerations for mobile contexts requiring their own special processes in the development cycle. Nonetheless, there are a number of challenges specific to mobile applications that set it apart from more traditional desktop-based applications. These include the diversity of technology, hardware constraints of battery life, memory, display, touch-screen interaction types, and their frequent use in outdoor environments. De Sá et al. (2008, p. 225) suggested 'new design approaches are required, particularly for the evaluation and prototyping phases'. A visual design process is proposed, one that is iterative and participatory, involving expert users and outdoor evaluations to address realistic settings. Eom and Lee (2013, p. 2) warned of the 'need to recognise the dynamically changing environment and properly react to those changes'. The rapid change of mobile technology and networks has significant impacts on approaches to application development, requiring software adaptation and much greater forward thinking and planning. Due to the special needs within the mobile computing environment, many argue that conventional software methodologies do not provide a full framework solution to mobile development (De Sá et al. 2008; Rahimian and Ramsin 2008; Eom and Lee 2013).

In the field of software development, three models are commonly employed in organising activities, time and people – the waterfall, iterative spiral, and agile software development models.

Waterfall model

The waterfall model (Figure 4.1) follows a series of cascading phases to be completed in a linear fashion. The phases start with development of requirements through to analysis, design, coding, testing, and then final operation and maintenance. Royce (1987) was the first to develop the concept of the waterfall model, although he did not use this name. Developed in 1970, it was put forward at a time when the software application specifications were generally developed by programmers with little input from a more broad set of stakeholders (Laplante and Neill 2004). The model was well suited to that context of little change in requirements and involvement of end-users. In today's conditions, the linear nature of the waterfall, requiring successive activities to be completed before the next can begin, makes it rigid and inflexible. It is therefore unsuitable for products that involve changing requirements or high complexity.

The waterfall model does have a number of advantages. It provides a simple, easy-to-understand approach and gives identifiable milestones. The waterfall model places emphasis on planning and documentation,

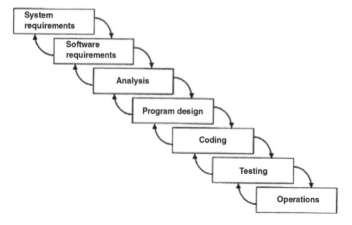

Figure 4.1 Waterfall model as described by Royce 1987.

which includes requirements and design. This is a worthwhile investment of time as it addresses issues at the early stages that could be more time-consuming and require much greater effort to fix if found later in the development cycle. For example, design flaws can be identified before they develop into something more serious. The intensive documentation through the process corresponds well to projects where quality control is an important concern.

The main disadvantage of the waterfall model is that it does not reflect the iterative nature of design and development. While the requirements phase may be thorough, it may not be accurate in predicting everything, particularly in relation to the needs of end-users. Royce (1987) described a model that allows for returning to an earlier phase, such as from design back to requirements. However, this can be costly. The difficulty and cost of changes increase substantially in proportion to the number of phases that have been completed. For example, if a new requirement is found during testing, a late stage in the project cycle, significant work will be required to swim upstream, thereby incurring substantial cost.

Spiral model

The spiral model (Figure 4.2), first put forward by Boehm in 1986, takes into account the iterative nature of design and development. Boehm (1986, p. 22) argued that the model encapsulates 'a risk driven approach to the software process, rather than a strictly specification-driven or prototype driven process'. The spiral model's radial dimension represents the cumulative nature of the steps and accommodates specification, prototype and simulated approaches. It aims to minimise risks by the repeated creation of prototypes that each undergo risk analysis. Each phase of the development, which outputs a new prototype, is made up of four stages: (1) planning and development of objectives, (2) identification and resolution of risks, (3) development and tests, and (4) evaluation and planning for next iteration. Subsequent spirals build on the previous one, with a prototype developed at the end of the risk-analysis phase. Each cycle is completed by a review by the primary people involved with the product. This allows the customer and end-users to evaluate the product before it continues to the next spiral. Once a satisfactory prototype design is reached, the software is constructed following the last three phases of the waterfall model – coding, testing, and then final operation and maintenance.

The model is flexible as to the amount of spirals to be included to reach a satisfactory design. The cumulative cost of the product is represented by the width of the spiral, measured from the epicentre to the outer edge of the last spiral. There is a potential high cost for the product output, tied to the number of people involved and number of prototypes produced. Compared to the waterfall model, the spiral approach is much more adaptable to a wide range of situations and change. It has the drawback of being much more labour-intensive. A key advantage of the model is the high degree of risk analysis it encompasses. This can be a disadvantage too as it requires specialist expertise, and the project's success is highly dependent on the risk-analysis phase. It is therefore most suitable for large projects rather than small ones with low budgets.

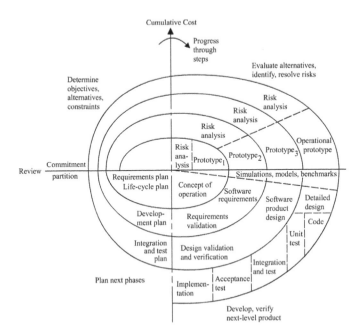

Figure 4.2 Spiral model as proposed by Boehm 1986.

Agile software development

Agile philosophy developed out of a desire to overcome the limitations of the earlier methodologies and make something more suitable for the dynamic environments of the modern times (Misra et al. 2012, p. 974). In 2001, a group of seventeen software practitioners developed the *Manifesto for Agile Software Development* (Fowler and Highsmith 2001). This manifesto specified a shift from traditional models to better define purpose and principles that encourage best practices in developing software. The statement of purpose declared value of people and interactions over processes and tools; emphasis on development of working software over significant documentation; customer collaboration; and a response to change. In addition to this statement of purpose, the manifesto also specified twelve principles to guide the approach.

The agile philosophy follows lightweight software development methods as compared to the heavyweight waterfall approach (Misra et al. 2012). A key focus is on collaboration, adaptive planning and rapid development. Its methodology employs iterative, incremental development processes extended from Boehm's spiral model, but does not place emphasis on following strict processes. A number of methodologies fall under the agile development practice such as Extreme Programming, Crystal Methodologies, SCRUM, Adaptive Software Development, Feature-Driven Development, and Dynamic Systems Development Methodology (Fowler and Highsmith 2001). Agile software development practice recognises that there is no single methodology that meets the needs of every situation and that processes can evolve over time.

Many within the software development industry have adopted agile philosophy, especially large private organisations and governments (Misra et al. 2012, p. 977). The principles of agile philosophy encourage simplicity of development and restricting to what is absolutely necessary. This accommodates change and corresponds to a key consideration within mobile-application development.

A challenge faced by developers of mobile apps is that the hardware and infrastructure constantly evolve in a fast-paced manner. The agile approach thereby gives a framework with which to plan and develop for the long term. The agile principles have provided a foundation for mobile computing developers to produce their own frameworks, taking into account the special needs of mobile computers (Rahimian and Ramsin 2008; Eom and Lee 2013). Rahimian and Ramsin (2008) recommended a hybrid methodology drawn from ideas of the Adaptive Software Development New Product Development. Eom and Lee (2013) proposed a human-centred software development framework that embeds agile principles, together with a process of agent-oriented software engineering and the agent technology support.

App development process and considerations

Regardless of the methodology employed for the development of a mobile application, the process will encapsulate common phases and deal with universal issues. This section looks at the key stages of the process and at considerations within each stage. This covers the make-up of the team, planning tasks, development, testing and deployment. While mobile application production has a unique set of requirements as compared to traditional desktop software, a good understanding of the broader aspects of project management is vital to ensure an efficient process and a successful end product.

Team formation

The size of the development team will naturally correspond to the scope and budget of the project. The make-up of the team should encompass a range of specialist personnel. This can help ensure a more successful product as expert knowledge improves efficiency and quality. A typical team for a mobile app would include the knowledge of industry experts, graphic designer(s), and programmer(s). Lee, Schneider and Schell (2004, p. 166) suggested that small teams of up to five developers should include a technical leader and several programmers. With large teams of up to twenty developers, the suggestion is to create a number of small teams with different areas of focus and each with their own technical leader. Each team could look after a different area of functionality. A single project manager could oversee the process.

Murphy (2011) differentiated mobile apps from conventional ICT projects by a number of factors, including that of the size of the development team being generally small and the deadlines typically short. In this light, developers will often release an app with a limited amount of features that will be expanded upon in a future update. This fits well with the iterative agile approach and highlights the importance of continued stakeholder feedback and evaluation.

Planning

All app developments generally begin with a planning stage to establish the goals and requirements. In this step of the process, a full evaluation is required about the app's purpose, the target audience and their specific needs. Defining and analysing the requirements involve gathering detailed information from a variety of sources through market research, end-user interviews, etc. Once the general requirements are defined, an analysis of the scope of development should be clearly articulated and documented. The scope document can form the basis of the agreement between the client and developer to avoid any dispute at a later stage. Table 4.2 illustrates common elements included in a scope document.

Table 4.2 Scope document form

Mobile App Scope Document	
Project Title:	
Project Manager:	
Date Prepared:	
Expected Start and Completion Date:	
Team Members: • People directly responsible for project deliverables and supporting team members • Project manager, programmers, graphic designer, knowledge expert	
Purpose of Project: • Goal statement • Benefits expected	
Background: • Brief project history • Justification for project • Unique qualities of project	
Deliverables: • Major outputs of the project • Measures of project success • Attach software requirements/feature specification document	
Stakeholders: • Client, end-user, other stakeholders of project	
Resource Requirements: • Identify life cycle requirements and costs such as implementation, ongoing production, purchases (including hardware and software) • Staff requirements and costs • Training needs and costs • Broad timeline, including project milestones	
Operations & Support: • Team/people responsible for knowledge base and documentation • Team/people responsible for product support • Team/people responsible for ownership and maintenance • Ongoing training needs • Communications and marketing requirements	

Risks and Contingency Plan:

- Risks related to people, budget, deliverables, timeline, support, and training
- Mitigation measures or contingency plans for issues or risks

Issues and Risks	Risk Mitigation or Contingency
1.	1.
2.	2.
3.	3.
4.	4.

In Rahimian and Ramsin's (2008) hybrid agile methodology, the process starts with idea generation followed by multiple steps of analysis – preliminary, business, and then a detailed analysis. The sequence through multiple forms of analysis aims to mitigate the risks within the project development.

Lee, Schneider and Schell (2004, p. 164) suggested the development of a project plan to aid with the organisation of the project. It would include all tasks and sub-tasks, organised by time, completion dates and associated people. The project plan can help ensure a more efficient production stage. It would also help better estimate the resources needed, including people, hardware and software, time and cost.

Design

The design phase visualises all the various requirements and project specifications to become an end product. Lee, Schneider and Schell (2004, p. 177) warned against over-thinking the design and thus ending up with 'analysis paralysis', which stalls a project, as well as under-design, which can lead to a flawed, insufficient application. Design considerations include users and usability as well as mobile-device types. The design process can include mock-up layouts or wireframe diagrams, which are outlines depicting the placement of text, images, and buttons. These visualise the composition of contents for the app display. A number of wireframing tools are available for creating mock-ups and clickable mobile app simulations. Examples include Axure, Balsamiq, Fluid, Justinmind, Mockflow, Moqups, Proto.io, Protoshare, and UXPin Wireframe. The tools, many of which are online cloud-based services, provide a flexible space and template structures to quickly create wireframes that draw from a library of device-specific buttons, menu items and other widgets. Figure 4.3 illustrates a wireframe mock-up of a mobile app interface using UXPIN.

Development

Development of apps has become easier with a number of tools available to minimise the coding required. These tools are of two categories: software

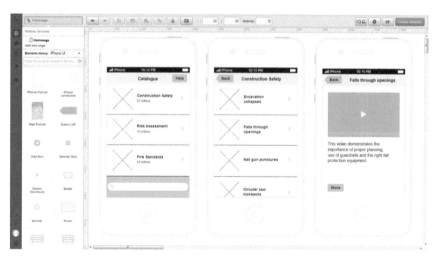

Figure 4.3 Wireframe mobile app mock-up using UXPIN.

Source: UserVoice 2013.

development kits (SDKs), and cross-platform application development plat-forms. The former category (see Table 4.3) is introduced by mobile operating systems (OS) companies and is native and restricted to the mobile devices produced by those companies. For example, Xcode was introduced by Apple for developers producing apps for its mobile OS. The latter cate-gory is flexible and apps can be developed for cross-platform deployment. Table 4.4 lists some of the widely used platforms in the app industry. PhoneGap, for instance, is one such tool for mobile app development. It is open source and enables publishing to a wide range of mobile operating systems, including iOS, Android, Windows Phone, Blackberry, Palm WebOS, and Symbian. Likewise, Canvas is both a tool and cloud-based service focused on data collection applications. As such, Canvas has been used to develop a number of mobile apps for the construction industry to facilitate on-site data collection. An example is TheContractorsGroup.com who have partnered with Canvas to develop apps that aim to reduce paperwork within the industry. The apps include Change Order, Estimate & Work Order, and Proposal.

Testing

Lee, Schneider and Schell (2004, p.183) suggested that testing be carried out from a number of perspectives, including functionality and usability, perfor-mance and security. A well-managed document process within testing should track changes and versions, and provide clear communication to all involved in order to avoid errors. Table 4.5 illustrates the key elements included in testing reports. This stage of app development should not be taken for

Table 4.3 Software development kits (SDK) direct from the mobile OS companies

Name	Description	Cost	Output compatibility
Android Open Source Project (Android 2013)	Java language, allows C, C++	Free	Android
BlackBerry Java Development Environment (JDE) (BlackBerry 2013)	Java language	Free	BlackBerry
Xcode (for iOS) (Apple 2013)	Objective-C or Objective Pascal	Only via app Store, needs review and approval by Apple Inc.	iOS
Windows Mobile (Microsoft 2013)	C, C++	Free	Windows Phone

Table 4.4 Application development tools

Name	Description	Cost	Output compatibility
Titanium (Appcelerator 2013)	Apps can be produced using web programming languages such as HTML, PHP, JavaScript, Ruby and Python	Free, but offer cloud service plans from US$999/month	iOS, Android, BlackBerry, Windows
AppMakr (PointAbout 2013)	Makes apps with web programming languages such as HTML, PHP, JavaScript, Ruby and Python	Free, but offer a US$79/month fee per app subscription for access to more-advanced features	iOS, Android, BlackBerry, Windows
Canvas (Canvas Solutions 2013)	Cloud based service and tool focused almost entirely on data collection applications.	Monthly (US$20) or Yearly (US$210) subscription or Pay As You Go	iOS, Android, Blackberry, Windows
JQuery Mobile (The jQuery Foundation 2013)	HTML5 based framework.	Free (Open source framework)	iOS, Android, Blackberry, Bada, Windows, webOS, Symbian, MeeGo
MoSync (MoSync AB 2013)	Development through C, C++, HTML5, CSS, JavaScript.	Free (Open source framework); subscription support services available from €199	iOS, Android, Blackberry, Java ME, Moblin, Symbian, Windows
M-Project (M-Project 2013)	JavaScript framework, HTML 5/CSS3 based features.	Free (Open source framework)	iOS Android, Blackberry, Windows

(continued)

Table 4.4 (continued)

Name	Description	Cost	Output compatibility
PhoneGap (Adobe Systems 2013)	Creates apps using HTML, CSS and Javascript	Free (Open source framework)	iOS, Android, Blackberry, webOS, Symbian, Bada, Windows
Sencha Touch (Sencha 2013)	Development through HTML5, CSS3 and JavaScript	Two versions available – free Open source licenses for application development, and a paid commercial license for OEM uses	iOS, Android, BlackBerry, Windows

granted. Insufficient time for thorough testing of the mobile application and the supporting infrastructure can result in key issues going unnoticed, thereby leading to a less reliable end product. Moreover, in iterative models, the testing step will occur on multiple occasions in the process.

Testing in mobile contexts provides a unique set of challenges, one of the biggest being the large range of operating systems and enormous range of devices. In addition, there are different flavours of the operating system

Table 4.5 Test report format

Test Report	
Project version	Version/build number of product
Device & platform	OS and device where issue appears
Issue type	Bug, New Feature, Cosmetic improvement, Usability
Severity	1. Cosmetic 2. Minor Failure 3. Major Failure 4. Crash/data loss
Issue title • Concise statement of issue	
Description • Detail description with steps to reproduce • Include actual result versus expected result	
Work log activity • Notes on steps in resolving issue	

specific to a particular device. For example, the Android flavour is different on a Samsung model as compared to that on an HTC. To ensure accurate results, testing should be carried out on the physical devices supported by the application. This is not always possible, particularly if a large range of devices are supported. Emulators, available as PC applications, provide an easy way to manage testing across the wide spectrum of mobile hardware. The problem is that emulators may not be 100% accurate and do not capture the full range of interactions that users will have with the product. It therefore makes sense to use a mix of real devices along with emulators and have a good plan of actions for app testing.

Deployment

While the submission of the completed application to an app distribution platform may seem to be the last step in the development process, it is typically the start of a new cycle of evaluation and redevelopment. The testing phase may include a degree of end-user review and evaluation. However, once the product is deployed, it is bound to undergo much greater scrutiny by end-users. It is worthwhile to have systems in place to capture problems and a means to address them both in the short and long terms. The new information from end-users, along with features out of the scope for the first release, could be revisited by returning to the planning stage. The market feedback, as well as plans to spread functionality across versions, can support continual product improvement and user engagement.

Developing a fall PtD mobile app for construction

The preceding sections described app development from a theoretical perspective. This section applies those concepts in developing a mobile app for fall PtD in construction. In a broader sense, the agile software development model was followed in this project due to its increased flexibility and cost-effectiveness as opposed to those of other models. The process involved and the outputs are described under two sections here, namely, app planning and modelling, and app building. The evaluation of the app is discussed in detail in the next chapter, with a separate section dealing with the testing of the app.

App planning and modelling

A fall PtD knowledge base was established through a detailed content analysis of PtD literature and industry reviews, as described in the previous chapter. The goal of the proposed app was to provide building designers (architects and engineers) with decision support during the early and detailed design stages to enable them to implement fall PtD effectively. Hence, it was crucial to understand and determine the expectations of potential users of the app in terms of its functionalities. Nonetheless, it would be difficult and ineffective to document user needs only by interviewing potential users when they have never had any experience with an app of this nature. Therefore, the author adopted an experiential approach to determine user requirements with the help of a rapid prototype/mock-up app.

The author developed a rapid prototype/mock-up app based on his ideas of the functionalities required for the app to support fall PtD decisions. Then, the mock-up was demonstrated to construction industry professionals, followed by interviews, seeking their approval for existing functionalities and suggestions for further functionalities. This enabled the identification of user requirements productively as the potential users experienced the proposed app through the prototype and responded pertinently. The demonstrations and interviews were held with architects and engineers in two well-known firms in Sydney; one of them was an architectural practice, while the other was a design–build contractor. Subsequently, the following key functional requirements were finalised for the app:

- the PtD knowledge base should be organised and presented in a modular fashion so that users can easily and quickly view/retrieve necessary information from the app;
- a content-filtering method should be made available in the app so that users would be able to obtain specific PtD suggestions for a particular design project undertaken at a given time;
- users should be able to save project-specific PtD suggestions for later use or for record-keeping purposes;
- facilities for the creation of PDFs of project-specific PtD suggestions, and the subsequent emailing of the PDFs, should be made available in the app;
- the app should be stand-alone and work without internet connectivity for improved applicability and flexibility in use at work, regardless of the user's location;
- from the author's perspective, it was necessary to include a disclaimer and acknowledgements for purposes of potential liability management.

Plainly understanding the app's functionality requirements led to the development of various conceptual designs for the app, as discussed in the forthcoming paragraphs.

In addressing the user needs of modularised PtD knowledge contents, a knowledge map was created to logically organise fall PtD knowledge into appropriate clusters. Altogether, there were 85 sets of design suggestions in the knowledge base, accompanied by 3D illustrations. These suggestions were arranged into clusters as outlined in Figure 4.4. The structuring of the knowledge map followed the natural phases and elements involved in building designs as well as the aspects to be considered in PtD design reviews.

The functionality for content filtering was achieved by the introduction of two questionnaires, corresponding to the two design phases shown in the knowledge map. Table 4.6 depicts the questionnaires in that the early-design PtD questionnaire has nine questions, while the detailed-design PtD

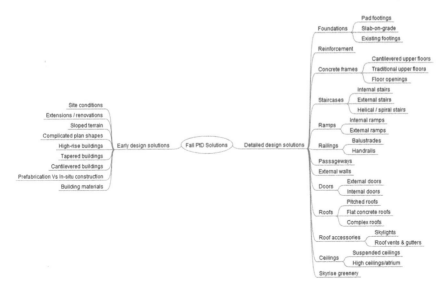

Figure 4.4 Fall PtD knowledge representation/map.

questionnaire has twenty-eight questions. The questions in the question-naires correspond to the clusters in the knowledge map. Completing the appropriate questionnaire at a given phase will result in the user obtaining project-specific PtD suggestions for the clusters.

It was decided that the app would have an in-built database to store PtD knowledge and project details, rather than adopting a server or cloud-based

Table 4.6 App questionnaires

Early Design PtD Questionnaire

Please answer the following questions to view project-specific, early design fall PtD suggestions.

1	Would you need to consider the impact of site factors on your design?	Yes	No
2	Is the proposed design for extension/renovation of an existing building?	Yes	No
3	Is the proposed building to be sited on sloped terrain?	Yes	No
4	Does the building design feature complicated plan shapes, such as multiple recesses, offsets or wavy sections, along the perimeter?	Yes	No
5	Is the proposed project a high-rise building?	Yes	No
6	Are you designing a tapered building?	Yes	No
7	Is any section of the proposed building cantilevered (this excludes usual balconies and small projected slabs, rather refers to an entire section of the building being cantilevered)?	Yes	No

---

(continued)

Table 4.6 (continued)

8	Would the building design require extensive in-situ concreting?	Yes	No
9	Would you need to specify building materials for the project?	Yes	No

Detailed Design PtD Questionnaire

Please answer the following questions to view project-specific, detailed design fall PtD suggestions.

1	Does the project design have isolated (e.g. pad) footings?	Yes	No
2	Is there a slab-on-grade in the design?	Yes	No
3	Are you introducing new footings near existing footings in a renovation/extension project?	Yes	No
4	Does the project involve extensive reinforcement design?	Yes	No
5	Are concrete frames designed for the project?	Yes	No
6	Are there cantilevered upper floors or roofs in the design?	Yes	No
7	Does your design require creating openings on upper floor slabs or roofs?	Yes	No
8	Are there humped, depressed or raised areas on your floor design?	Yes	No
9	Are staircases designed for the project?	Yes	No
10	Is there a spiral or helical staircase in the design?	Yes	No
11	Does the project have external staircases?	Yes	No
12	Is any ramp designed for the project?	Yes	No
13	Does the project have an external ramp?	Yes	No
14	Are there passageways or corridors in the design?	Yes	No
15	Are railings required for the project?	Yes	No
16	Does the project design specify external walls on upper floors?	Yes	No
17	Does the design feature external doors on upper floors or swing doors?	Yes	No
18	Are there doors near staircases in the design?	Yes	No
19	Does the design require upper floor windows?	Yes	No
20	Would you need to specify floor finishes for the project?	Yes	No
21	Is pitched roofing involved in the project?	Yes	No
22	Is a concrete flat roof designed for the project?	Yes	No
23	Does the design specify complex-shaped roofs?	Yes	No
24	Are skylights included in the roof design?	Yes	No
25	Are roof vents or gutter systems specified in the roof design?	Yes	No
26	Are you designing suspended ceilings for the project?	Yes	No
27	Is there an atrium or high ceiling in the project design?	Yes	No
28	Are you designing skyrise greenery systems for this building (e.g. rooftop vegetation or wall/ledge planting)?	Yes	No

model. User operations and functionalities of the app were modelled as shown in Figure 4.5, which lays out the app pages and explains how different functionalities are integrated into different pages. The key functions of the app are as follows.

1   **Splash screen** – will be shown on app launch. After a few seconds, the application will start.

2   **Disclaimer** – will be shown after the splash screen, displaying disclaimer contents. The user will have to accept the disclaimer to use the app.

3 **Acknowledgements** – Once the user has accepted the disclaimer, the user will be navigated to the acknowledgement page, which displays predefined contents, and the user will be able to use the app for decision making after navigating from this screen.

4 **Home screen** – will act as the main screen of the app and will provide two options for the user. Available options will be: early-design fall PtD, and detailed-design fall PtD. The user can select a preferred option.

5 **Early/detailed design fall PtD screen** – displays informative text and three options for the user, namely, all PtD suggestions, project-specific PtD suggestions, and saved projects.

6 **All PtD suggestions** – clicking on this will enable the user to view all suggestions stored in the app under the category (either early design or detailed design). Knowledge contents for the categories are arranged according to the clusters shown in the knowledge map. In order to simplify knowledge retrieval, the PtD suggestion sets and the associated 3D illustrations for a given cluster are laid out in a thumbnail fashion, in which, whenever the user selects a thumbnail, it will expand in a separate page and display the full content of PtD suggestions. This also offers improved flexibility and ease in moving across different sections and subsections.

7 **Project-specific PtD suggestions** – this option allows the user to create new projects wherein the user can view suggestions based on their project requirements. By selecting this option, the user will be asked to provide a project name to start. Thereafter, the user will be presented with a questionnaire, which the user would need to complete to obtain suggestions tailored to their project needs.

The questionnaire presents predefined early/detailed-design fall PtD questions wherein the user has to answer each question by selecting either Yes or No. Answers provided by the user will be immediately saved. The user has to answer all the questions to view the suggestions, and at least one question must be answered to have the project saved in the app for future retrieval.

The user would have an option to reset the answers. It will clear answers of all the questions for a given project. However, a project without any answers will not be saved. Hence, a reset project will not be saved. If a project's answers are reset and no questions are answered thereafter, then the previous answers in the project will be retained.

Upon completing the questionnaire, the user will be presented with the suggestion page whereon the user can select any single suggestion and view, generate a PDF of all project-specific suggestions, or share the PDF via email.

8 **Saved projects** – previously saved projects will be listed on this screen where the user can view suggestions, change answers to the questionnaire, or delete the project from the database.

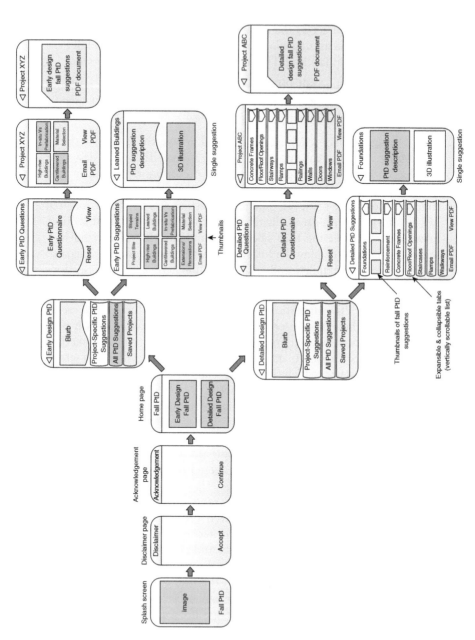

Figure 4.5 App layout model.

Following the planning and scoping process, wireframes for the app were created using OmniGraffle, based on the aforementioned functional requirements, business logics and view layouts. These wireframes then guided the programmer in building the app, which is discussed in the next section.

App building

The proposed fall PtD mobile app was developed to be compatible with Apple mobile devices such as iPhones and iPads. Mobile apps can be developed for a single or multiple platform compatibility, such as Apple, Android, and Windows mobile. However, building mobile apps for multiple platform deployments requires cross-platform testing, additional resources and extra time. In order to optimise the efforts and resources in the research, it was decided to build only an Apple-compatible app.

Apple's SDK, Xcode, was fully utilised to develop the fall PtD mobile app. Xcode is an Integrated Development Environment (IDE), which unifies user interface design, coding, testing and debugging within a single window to make app development tasks easier, faster and more effective. Being an object-C programming environment, Xcode deploys the Model-View-Controller (MVC) architecture. Figure 4.6 displays the MVC design for the fall PtD mobile app in that the 'fall PtD model' houses the database/knowledge base of the app, and the 'fall PtD view' lays out the interfaces,

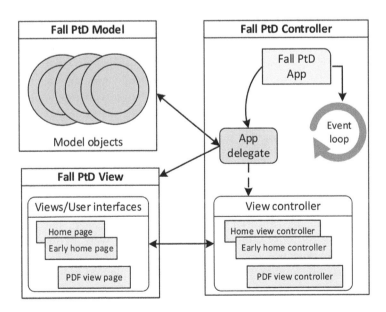

Figure 4.6 MVC architecture of the app.

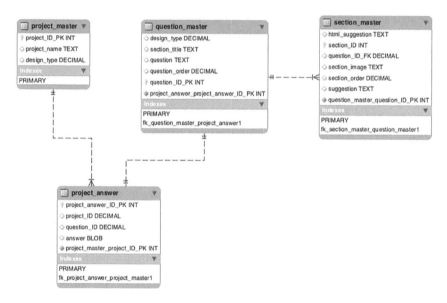

Figure 4.7 Entity-relationship diagram.

whilst the 'fall PtD controller' connects the interfaces with the database/ knowledge base. The roles and operations of these different architectural components are described in the subsequent paragraphs.

The model objects represent and hold knowledge and expertise related to fall PtD as well as define logics that manipulate the knowledge for PtD decision support. The app and its use for PtD decision support are driven by the PtD knowledge base. Effectively organising and implementing the knowledge holders and manipulation logics were therefore critical to the error-free operation of the app. An entity-relationship diagram was first created for this purpose, as shown in Figure 4.7, which was then translated into a relational database in Xcode. The data table 'section_master' is the repository of fall PtD knowledge, and the table 'project_master' is the holder of project details. Projects are connected to the PtD knowledge content through the table 'question_master' in generating outputs of project-specific PtD suggestions, based on answers provided by app users to the questions, which are captured in the table 'project_answer'. The logics of manipulating the data contained in these tables were defined as queries in the model objects.

The view objects are in charge of retrieving and displaying model contents according to user requests, as well as housing different user request options such as creating project-specific fall PtD suggestions, viewing all early-design PtD suggestions, emailing PDFs of PtD suggestions, etc. Figure 4.8 illustrates the organisation of different views/pages in the app as

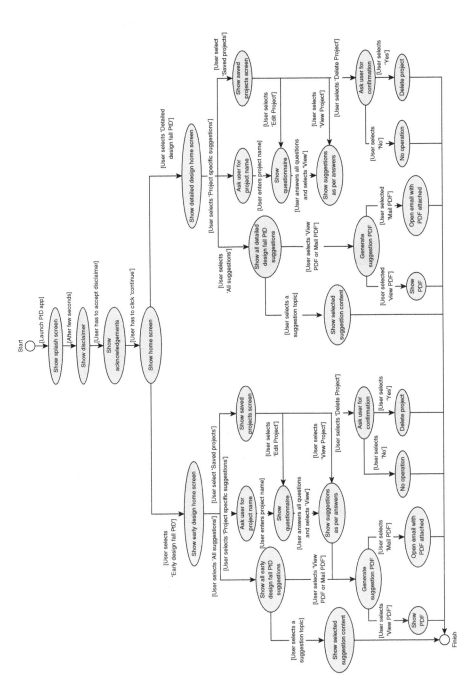

Figure 4.8 App operation model.

Figure 4.9 App screen examples.

well as how a user would interact with the app and what operations a user could, or might need to, perform. The views were physically constructed in Interface Builder of Xcode. Figure 4.9 shows iPhone versions of some screenshots of the app views. The first three images show a user's path to reach the app's home screen. The second three images depict how project-specific PtD suggestions are obtainable at the early-design phase, whilst the last three images illustrate how individual suggestions from the 'all PtD suggestions' can be viewed.

The controller objects act as intermediaries between the views and the models, and play the role of handling requests from users. In essence, they dictate what to do behind the scene in the model objects and what next to display in the view for requests made by users. There are controller objects corresponding to each view object in the app. For example, Home view controller for Home screen, PDF view controller for PDF views and so on. When an action is triggered in a view object by a user, the controller receives the request, decides the requested activity, delegates tasks to be performed in the model objects for the request, and then sends commands to its associated view to change its presentation.

Lessons learnt

Several valuable lessons were learnt during the development process of the new fall PtD mobile app. Some noteworthy ones are reflected upon below, which may be beneficial for other researchers who intend to build mobile apps for supporting similar construction management issues.

Establishing a clear scope, terms of reference and conceptual models before starting the development of the app helped with an accelerated building of a working app without major technical or conceptual issues. The use of a rapid prototype for user-requirement identification and the early involvement of the programmer/developer in discussions enabled the researcher to produce effective terms of reference and conceptual models for the app. They also reduced possible coding hiccups caused by erroneous conceptual models. These were crucial given that the time and budget available for the app development were limited. Iterations in the process were unavoidable, but were not complicated in the way they might have been if the scope had not been well-defined.

The production of 3D illustrations for the PtD knowledge base consumed an extended time, but they enhanced the readability and applicability of the PtD suggestions offered by the mobile app. However, it posed some technical challenges, including: (1) the app size was rather large, in the range of 100MB, which slowed down installation and caused errors, and (2) the PDF files created out of the app were too large to be emailed by the mobile device's native email application. This issue was overcome by compressing the image file sizes. Hence, when 3D images and illustrations are used in app developments, pre-considerations are important to decide the allowable file size for images. Even if the images may be stored in a server or cloud that an app refers to for image display, output generation and sharing may be restricted.

Another important lesson learnt was the need to follow the app distributor's terms and conditions – in this case, it was Apple's terms and conditions. Apple, as an app distributor, has set out conditions that developers must follow when developing apps for distribution through Apple Store. These conditions may conflict with certain functionalities planned by an app developer. In this development, for example, it was initially decided that the app would close programmatically if the user chooses to decline the disclaimer. However, programmatically closing apps was not allowed by Apple conditions, which resulted in minor modifications to the app. Hence, it is important that app developers consider app distributors' terms and conditions at the time of scope and concept planning.

Equally important is to check interoperability if the newly developed app is to interface with other apps. For example, in this app, it was necessary to ensure that the PDF created by the app did not exceed the maximum allowable attachment by the native email applications of the targeted devices.

Conclusion

Even though mobile computing devices continue to expand and the amount of mobile applications available grows exponentially, the use of this technology for accident PtD in construction remains very limited. This chapter discussed the development of a mobile app for fall PtD in an endeavour to fill this knowledge gap. The development process followed the logical system-development order and produced a working app, which is able to offer building designers (architects and engineers) mobility and flexibility in curtailing the number of falls on construction sites through their safe designs. The next chapter is devoted to discussing a comprehensive performance evaluation of the new fall PtD mobile app from both technical and business perspectives.

References

Aconex. (2013) *Aconex Mobile*. URL (accessed 17 June 2013): http://www.aconex.com/mobile-iphone.

Adobe Systems. (2013) *PhoneGa*. URL (accessed 17 June 2013): http://phonegap.com.

Ahsan S, El-Hamalawi A, Bouchlaghem D and Ahmad S. (2007) Mobile technologies for improved collaboration on construction sites. *Architectural Engineering and Design Management*, 3(4): 257–272.

a la mode. (2013) *FormMobi*. URL (accessed 17 June 2013): https://www.formmobi.com.

Android Open Source Project. (2013) *Android SDK*. URL (accessed 17 June 2013): http://developer.android.com/sdk/

Anumba C J and Wang X. (2012) Mobile and pervasive computing in construction: An introduction. In: Anumba C J and Wang X (eds), *Mobile and Pervasive Computing in Construction, First Edition*. Oxford: John Wiley & Sons, Ltd, pp. 1–10.

Appcelerator. (2013) *Titanium*. URL (accessed 17 June 2013): http://www.appcelerator.com/platform/titanium-platform/.

Apple. (2013) *Xcode*. URL (accessed 17 June 2013): https://developer.apple.com/xcode/.

Auman Software. (2013) *sightLevel Pro*. URL (accessed 17 June 2013): http://www.isightlevel.com.

Autodesk. (2013a) *AutoCAD 360*. URL (accessed 17 June 2013): https://www.autocad360.com/products/mobile/.

Autodesk. (2013b) *Buzzsaw*. URL (accessed 17 June 2013): http://usa.autodesk.com/buzzsaw/.

BlackBerry. (2013) *BlackBerry Java Development Environment*. URL (accessed 17 June 2013): http://developer.blackberry.com/java/download/.

Boehm, B W. (1986) A spiral model of software development and enhancement, *ACM SIGSOFT Software Engineering Notes*, 11(4): 21–42.

Canvas Solutions. (2013) *Canvas*. URL (accessed 17 June 2013): http://www.gocanvas.com.

Chen Y and Kamara J M. (2008) Using mobile computing for construction site information management. *Engineering, Construction and Architectural Management*, 15(1): 7–20.

Deere J. (2013) *JDLink*. URL (accessed 17 June 2013): http://www.deere.com.au/wps/dcom/en_AU/products/equipment/agricultural_management_solutions/information_management/jdlink/jdlink.page.

De Sá M, Carriço L, Duarte L and Reis T. (2008) A mixed-fidelity prototyping tool for mobile devices, *Proceedings of the working conference on Advanced visual interfaces* (AVI '08), ACM, New York, NY, USA, pp. 225–232.

DeWALT. (2013) *DeWALT Mobile Pro*. URL (accessed 17 June 2013): http://www.dewalt.cengage.com/mobilepro/.

Double Dog Studios. (2013) *Home Builder Pro Calcs*. URL (accessed 17 June 2013): http://doubledogstudios.com/apps/homebuilderprocalcs/.

Eom H and Lee S. (2013) Human-centered software development methodology in mobile computing environment: agent-supported agile approach. *EURASIP Journal on Wireless Communications and Networking*, Springer, 2013(1): 1–16.

Fowler M and Highsmith J. (2001) The agile manifesto. *Software Development Magazine*, 28–32.

Graphisoft. (2013) *BIMx*. URL (accessed 17 June 2013): http://www.graphisoft.com/bimx/.

Idan Sheetrit. (2013) *i-Ruler*. URL (accessed 17 June 2013): https://itunes.apple.com/us/app/i-ruler/id474785950?mt=8.

IMSI/Design LLC. (2013) *TurboViewer*. URL (accessed 17 June 2013): http://www.turboapps.com/TurboApps/TurboViewerPro/tabid/2224/Default.aspx.

JBKnowledge Technologies. (2013) *SmartBidNet*. URL (accessed 17 June 2013): http://smartbidnet.com.

Koseoglu O O and Nielsen Y. (2005) Mobile computing in construction projects; assessing feasibility, *Proceedings of the 21st Annual ARCOM Conference*, 7–9 September 2005, London, UK, pp. 561–570.

Laplante P A and Neill C J. (2004) The demise of the waterfall model is imminent. *Queue*, 1(10): 10–15.

Lee V, Schneider H and Schell R. (2004) *Mobile Applications: Architecture, Design, and Development.* Upper Saddle River, NJ: Prentice Hall.

Loupe. (2013) *PlanGrid.* URL (accessed 17 June 2013): http://www.plangrid.com.

Maxwell Systems. (2013) *ProContractorMX Mobile Connect.* URL (accessed 17 June 2013): http://www.maxwellsystems.com/products/construction/procontractormx/modules/mobile-connect.

Microsoft. (2013) *Windows Phone Software Development Kit.* URL (accessed 17 June 2013): http://www.microsoft.com/en-au/download/details.aspx?id=27570.

Misra S, Kumar V, Kumar U, Fantazy K and Akhter M. (2012) Agile software development practices: Evolution, principles, and criticisms. *International Journal of Quality & Reliability Management,* 29(9): 972–980.

MoSync AB. (2013) *MoSync.* URL (accessed 17 June 2013): http://www.mosync.com.

M-Project. (2013) *M-Project.* URL (accessed 17 June 2013): http://www.the-m-project.org.

Murphy C. (2011) App dev must get agile enough for mobile. (Global CIO) (Column). *InformationWeek,* no. 1313, p. 10.

PointAbout. (2013) *AppMakr.* URL (accessed 17 June 2013): http://www.appmakr.com.

Rahimian V and Ramsin R. (2008) *Designing an Agile Methodology for Mobile Software Development: A Hybrid Method Engineering Approach.* URL (accessed 30 Sept. 2014): http://ieeexplore.ieee.org/stamp/stamp.jsp?tp=&arnumber=4632123.

Royce W. (1987) Managing the development of large software systems, *Proceedings of the 9th International Conference on Software Engineering,* Monterey, California, USA: IEEE Computer Society Press, pp. 328–338 (reprinted from Proceedings of IEE WESCON, August 1970, The Institute of Electrical and Electronic Engineers Inc, TRW, pp. 1–9).

Sencha. (2013) *Sencha Touch.* URL (accessed 17 June 2013): http://www.sencha.com/products/touch.

The jQuery Foundation. (2013) *JQuery Mobile.* URL (accessed 17 June 2013): http://jquerymobile.com.

True Context. (2013) *Prontoforms.* URL (accessed 17 June 2013): http://www.prontoforms.com.

UDA Technologies. (2013) *Onsite Planroom.* URL (accessed 17 June 2013): http://www.prontoforms.com.

UserVoice. (2013) *UXPIN.* URL (accessed 14 June 2013): http://uxpin.com.

5 Fall Prevention through Design mobile app evaluation

Introduction

This chapter discusses the evaluation process of the newly developed fall Prevention through Design (PtD) mobile app and its outcomes. The aim of the exercise was to assess: (1) how effective the mobile app is for fall prevention in construction, and (2) if any technical errors are encountered during the use of it. First, an app evaluation framework is developed, encapsulating comprehensive assessment criteria for an effective process. Then, the administration of a case-study-driven charrette workshop to evaluate the app in accordance with the framework is explained. Finally, the evaluation findings are expounded and reflected upon.

Mobile app evaluation framework

The evaluation of the performance of a newly developed information system is a critical stage in the system-development cycle. ICT literature states that system evaluations comprise both verification and validation. Sojda (2007) explained these terms as follows:

- *verification* – the exercise of ensuring from a system modelling and programming standpoint that an information system is internally complete, coherent and logical; in simple terms, it is making sure that decision logics, algorithms, databases, knowledge bases and other structures have been coded correctly;
- *validation* – the exercise of ensuring that a newly developed system addresses its intended purpose and is helpful to the end-user for making better decisions, avoiding poor decisions, or making decisions more quickly with a minimum need of looking for information/ knowledge elsewhere.

Dařena (2011) suggested that system evaluations need to consider technical and technological performances as well as impacts on management of the organisational processes and cost. It was therefore essential to adopt a holistic framework in order to ensure an effective evaluation of the mobile app was carried out. Recently emerged information system evaluation frameworks were studied and a new, suitable framework for the mobile app evaluation was then synthesised.

Criteria for evaluation

Heo and Han (2003) proposed six variables for measuring system performance, namely:

- system quality – operational efficiency of system functions, responsiveness of hardware and software, reliability, and ease of use;
- information quality – accuracy of information, timeliness of information, and completeness of information;
- information use – frequency of use, and voluntary usage;
- user satisfaction – attitudes about system function, and perceived utility of the system;
- individual impact – improved efficiency, and improved communication and teamwork;
- organisational impact – cost savings/improved return, and improved customer service.

Similarly, Yusof, Paul and Stergioulas (2005) introduced the '*why, what, who, how and when*' approach, which guides system evaluations through five reflective questions:

- Why is the evaluation being done?
- What is being evaluated?
- Who affects the evaluation and how?
- How is the evaluation to be carried out?
- When is the period of evaluation?

The *why* factor relates to the purpose of system evaluation, which can often be the appraisal of value, measure of success, or recognition of benefits. For the *what* factor, Serafeimidis and Smithson (2000) argued that system evaluations should see beyond narrowly direct appraisals of costs and benefits and include a study of intangible benefits, risks and opportunities presented by an information system. Stockdale and Standing (2006) claimed that there could be four categories of stakeholder for a system evaluation, namely, initiators of the evaluation, the evaluators who conduct the evaluation, the users of the system being evaluated, and a range of other parties such as IT managers, workers affected by change, etc. The evaluator should decide which stakeholders are relevant to a given system evaluation. Jamo Solutions (2013) stressed that user satisfaction with a mobile application is an important performance criterion. Hence, users or potential users are a key stakeholder for evaluations. The *how* factor includes the conduct of the evaluation with appropriate methodologies and instruments such as simulation modelling,

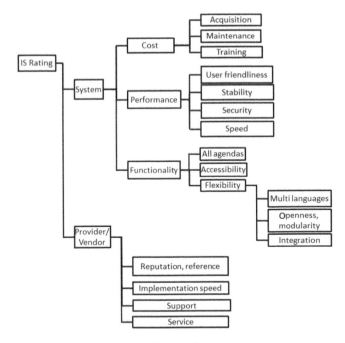

Figure 5.1 Dařena's IS evaluation framework.

Source: Dařena 2011.

cost-benefit analysis, return on investment, case study, or the measure of user satisfaction (Stockdale and Standing 2006). The final *when* factor dictates whether to treat the evaluation as a summative exercise at one point or as a continual exercise through different stages of a system implementation.

In a structured manner, Dařena (2011) developed a hierarchy of criteria for system evaluation, as shown in Figure 5.1, which can help answer 'what is being evaluated?' in Yusof, Paul and Stergioulas (2005)'s *why, what, who, how and when* approach. The hierarchy has two major branches: system, and provider/vendor, with associated sub-branches. The 'system' branch evaluates the information system as a core product by assessing its cost, performance and functionality, whilst the 'provider/vendor' branch evaluates the information system supplier's reputation, and speed of delivery and service/support.

Synthesis of an evaluation framework for the mobile app

Having reflected upon the various criteria proposed above for system evaluation and the context of the app use, a suitable framework was developed for evaluating the fall PtD mobile app. Figure 5.2 illustrates the four criteria adopted for the framework, namely: (1) quality of information provided by the app (completeness, usefulness), (2) functionality of the app

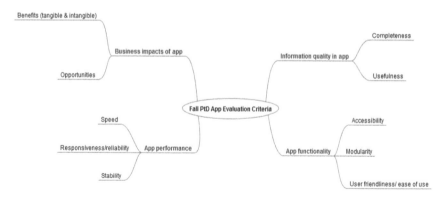

Benefits (tangible & intangible)

Business impacts of app

Opportunities

Completeness

Information quality in app

Usefulness

Fall PtD App Evaluation Criteria

Speed

Responsiveness/reliability   App performance

Stability

Accessibility

App functionality   Modularity

User friendliness/ ease of use

Figure 5.2 Fall PtD app evaluation framework.

(accessibility, modularity, user-friendliness/ease of use), (3) performance of the app (speed, responsiveness/reliability, stability), and (4) business impacts presented by the app (benefits, opportunities). This framework dictated what needed to be evaluated in the mobile app. Nonetheless, how one can conduct the evaluation based on this framework is still a consideration. The next section elaborates on the selection of a suitable technique.

Methods for system evaluation

Kamardeen (2009) and Sojda (2007) discussed four different approaches to system evaluation, namely, gold standard, field test, Turing test (also known as panel of experts), and predictive evaluation. The differences in the application of these approaches are as follows.

- In the **gold standard** approach, system performance is tested against test cases with known outcomes. The test cases can be developed from real-world scenarios or based on expert opinions.
- When adopting the **field test** method, the system is placed in the field that it is meant for, and its performance and errors are observed as they occur.
- In the **Turing test (panel of experts)** model, the performance of a system is tested against an independent panel of domain experts who were not involved in the system development process. Cases would be presented to both the panel of experts and the system independently for solutions. Then, the outputs of the system would be compared against the decisions of the panel. This is a common system-evaluation technique in the field of artificial intelligence.
- The **predictive evaluation** approach requires the use of historic data sets with known results. This approach is largely used for data-driven

decision support system (DSS) evaluations. The DSS is driven by past input data from a historic data set and the outputs are compared with the corresponding results in the data set. Generally, when a data-driven DSS is developed, the data set is divided into two independent parts: one for DSS development, and the other for DSS evaluation.

The selection of an appropriate evaluation technique is largely influenced by the type of system and the nature of the work that the system has been developed to support, as well as the extent of evaluation needed. The gold standard, the Turing test and the predictive evaluation approaches are more focused on the validation aspect of system evaluation, whereas the field test approach can be suitable for conducting a complete evaluation. This study, drawing from the notion of the field test, adopted a case-study-driven charrette workshop technique of evaluation. Charrette workshops allow conflicting issues around proposed design solutions to be tabled and evaluated by stakeholders, leading to the arrival of refined design solutions in an efficient and collaborative manner (Smith 2012). This is a well-established methodology for design reviews in the construction industry, whether they are for project-feasibility studies, value-management studies or safe-design reviews. This approach was deemed more appropriate for the evaluation of the new mobile app because:

- the process exercised in the evaluation would be a simulation of how the mobile app would be used in the industry in different projects/scenarios;
- it would provide a quicker evaluation opportunity;
- feedback from the evaluation participants is drawn from first-hand experience with the process and the app, thus making it more effective.

The next section describes the evaluation exercise and outcomes in vivid detail.

Fall PtD app evaluation process

The case-study-driven charrette workshop that evaluated the fall PtD mobile app was administered with eight early-career building professionals in a session that lasted for about four hours. The participants were identified using the widely accepted stakeholder-analysis technique classically developed by Freeman (1984), which categorises stakeholders according to their salience. The participant mix included five architects, two civil engineers and one builder. All of them possessed at least a bachelor's degree in their relevant discipline.

On commencement of the workshop, a short questionnaire survey was conducted with the participants to assess the degree of their pre-knowledge, skills and experiences in accident PtD. Responses to the following four questions were sought for this purpose:

- *Your experience in building design:*

  ☐ *Less than*   ☐ *1 to*   ☐ *3 to*   ☐ *5 to*   ☐ *More than*
      1 year   *3 years*   *5 years*   *10 years*   *10 years*

- *Your experience in design reviews for accident Prevention through Design:*

  ☐ *Less than*   ☐ *1 to*   ☐ *3 to*   ☐ *5 to*   ☐ *More than*
      1 year   *3 years*   *5 years*   *10 years*   *10 years*

- *Have you attended any skill development courses/seminars/workshops before on accident Prevention through Design principles (also known as safe design)?*

  ☐ *Yes*   ☐ *No*

- *How would you rate your current level of skills/knowledge in making better design choices to prevent fall accidents during construction and maintenance of buildings (1 = low; 5 = high)?*

  1   *2*   *3*   *4*   *5*

The responses provided by the participants are summarised in Figure 5.3 with four bar charts. In each chart, the responses are spread out along the X-axis, and the frequency of the selection of a response by the participants is represented in the Y-axis. These graphs demonstrate that the participants possessed only limited knowledge, skills and experience in fall accident PtD before attending the workshop.

Upon completion of the initial questionnaire survey, the workshop proceeded in the following manner:

- a PowerPoint presentation covering the concept of accident PtD, risk-assessment matrix and risk-control hierarchy was given to the participants, which lasted for about 30 minutes;
- then, the fall PtD mobile app was introduced, and the participants installed it on their iPhones or iPads;
- next, two groups were formed and were assigned the task of reviewing two real building designs with the support of the mobile app; one of them was a conceptual design, while the other was a detailed design;
- the design-review exercise lasted for about three hours and the participants recorded the review results in a template provided.

The ensuing sections explain the case-study projects and the review results.

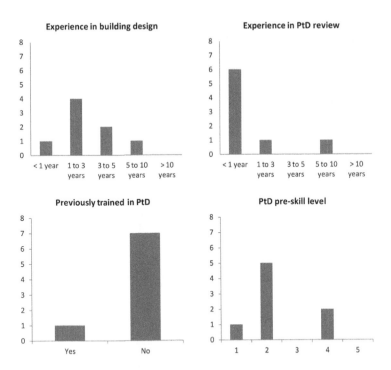

Figure 5.3 Profile of workshop participants.

Review of Case 1: conceptual design of a commercial building

This fall PtD review was conducted for the conceptual design of a commercial building, which consists of six levels, including a basement, and a footprint area of 22 m x 100 m. Facades along the east and west sides of the building are tilted, as shown in Figure 5.4, which results in a sloped concrete rooftop. While the basement of the building serves as parking space, the ground floor is designated as lobby area with informal meeting areas for building occupants and their visitors. It also contains exclusive spaces for retail and for food and beverages. Other floors house largely offices and some retail spaces. A roof terrace is positioned as a relaxation area at the points of the building where the tilting facades become vertical before tilting in the opposite direction. The building is serviced by a lift located in the middle, a staircase and two fire-exit stairs on the east side, as illustrated in the floor plan in Figure 5.5. While the frame of the building is constructed of in-situ concrete, all the facades are predominantly glazed. Internal office spaces are either kept open or glass partitioned. Toilet areas and fire-exit staircases are enclosed by brick walls.

The participant group analysed the design and discovered, within the limited time available, eight critical design issues that are conducive to fall hazards. Table 5.1 represents the findings of the design review. The first column of the table identifies the design elements/components analysed, while the second column details the fall hazard inherent in the design. In the third column, a risk rating is provided for the hazard according to the risk-assessment matrix shown in Figure 5.6. The fourth column records participants' suggestions for design revision, which were derived from their understanding of the PtD suggestions in the mobile app. In the final column, the type of risk control that the suggestion belongs to is identified, which is either elimination, substitution, isolation, engineering control, administrative control, or personal protective equipment. Figure 5.7 provides detailed explanations for these various risk-control techniques.

Figure 5.4 Perspective view of case study building 1.

(Image courtesy of Ganellen.)

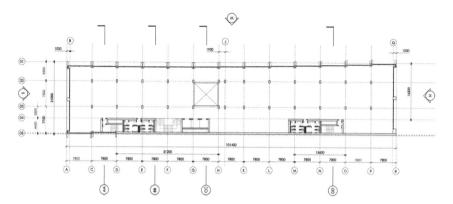

Figure 5.5 Third-floor plan of case study building 1.

(Image courtesy of Ganellen.)

Table 5.1 PtD review results for case study building 1

Design aspect analysed	Details of fall hazards identified	Risk assessment	Prevention through Design suggestion by the review panel	Type of risk control applied
Building frame	Concreting for exterior facade bays on the east and west sides are angled, creating fall hazards during construction	Likelihood: *Possible* Impact: *Major* Risk rating: *Very high*	Consider replacing in-situ concrete with prefabricated elements, particularly for the tilted facade bays	Substitution
Tilted building sides – east and west facades	Setting up scaffolding for the facade installation is difficult, which may cause falls	Likelihood: *Almost certain* Impact: *Catastrophic* Risk rating: *Very high*	Design-in strong scaffold connection points and sockets along the perimeter of the structural elements to brace and continue scaffold to make the required working height safe	Engineering
Glass facades – east and west sides of the building	Angles change across panels, which can cause difficulties in both installation and maintenance	Likelihood: *Likely* Impact: *Major* Risk rating: *Very high*	Add permanent anchor bolts to vertical building elements to provide lifeline tie-off points during installation and maintenance	Engineering
Reflective glass panels for facades	Cleaning and maintenance have inherent dangers through glare reflection, possibly resulting in loss of balance and falls	Likelihood: *Likely* Impact: *Major* Risk rating: *Very high*	1. Use non-reflective glass panels 2. Design for inclusion of shading devices such as external, horizontal louvres	1. Substitution 2. Engineering
Glass façades	Regular maintenance and cleaning of glass facades at heights presents fall risks to maintenance workers	Likelihood: *Possible* Impact: *Catastrophic* Risk rating: *Very high*	Reduce the placement of windows and glass facades that require regular maintenance and cleaning	Elimination

(continued)

Table 5.1 (continued)

Design aspect analysed	Details of fall hazards identified	Risk assessment	Prevention through Design suggestion by the review panel	Type of risk control applied
No protection/ barrier along inside edges of glass facades	If the window/panel breaks, users may fall out of the building	Likelihood: *Possible* Impact: *Catastrophic* Risk rating: *Very high*	1. Use high-strength windows and glazing systems to withstand accidental and casual loading 2. Design a railing system along the inside edges of facades	1. Substitution 2. Isolation
Lift shaft and floor openings for stair wells	Risk exists for construction workers to fall through openings	Likelihood: *Possible* Impact: *Catastrophic* Risk rating: *Very high*	Continue one layer of reinforcement bars through floor and lift shaft openings that can be cut and removed after construction	Engineering
The pitch of the roof slab is steep (close to 30°) in some places	Slip and fall accidents can occur during construction and maintenance of the roof	Likelihood: *Possible* Impact: *Catastrophic* Risk rating: *Very high*	1. Reduce building angle rotation so that the roof pitch is lower than 30° 2. Flatten lower part of the roof towards the edge so falls will be halted prior to the end of the building 3. Add permanent anchorage points on rooftop to provide tie-off points during construction and maintenance	1. Substitution 2. Substitution 3. Engineering

Likelihood of incident occurring (How likely is the incident to happen?)	Degree of Impact/Damage/Consequences (How severe will the incident hurt someone if it happens?)				
	Catastrophic (death or permanent incapacities)	Major (extensive injuries/long-term incapacities)	Moderate (injuries requiring medical treatments)	Minor (injuries requiring minor/first-aid treatments)	Negligible (no injuries)
Almost certain (expected in most circumstances)	Very high	Very high	Very high	High	High
Likely (will occur in most circumstances)	Very high	Very high	High	High	Moderate
Possible (might occur at some time)	Very high	Very high	High	Moderate	Low
Unlikely (could occur at some time)	Very high	High	Moderate	Low	Low
Rare (may occur in exceptional circumstances)	High	High	Moderate	Low	Low
	Risk Rating				

Figure 5.6 Risk-assessment matrix.

(Adapted from AS/NZS 4360, 1999.)

Control	Effectiveness	Description	Effort required
Start here ► Elimination	100% Hazard removed	Remove or design out the hazard e.g. removing extended cantilevers in a building design	Minor
Substitution	75% Hazard reduced significantly	Hazard substituted with something of lesser risk e.g. specifying a steel post instead of timber for kids' playgrounds to reduce the risk of splintering; specifying precast slabs instead of in-situ to remove hazards caused by formwork	Low
Isolation	50% Hazard reduced with isolation/confinement strategies	Hazard controlled through isolation e.g. isolating hazardous plant/chemicals from people; building permanent barriers around hazardous areas to keep people away	Moderate
Engineering	50% You are reducing the hazard with engineering solutions	Hazard controlled through engineering e.g. installing mechanical ventilators in a confined workspace	Moderate
Administration/ Training	25% You are introducing soft controls that rely on people following procedures	Communicating remaining risks; hazards controlled by influencing people e.g. signage, warnings, site inductions, etc.	High
Personal Protective Equipment (PPE)	5% You are limiting the damage	Provide people with safety equipment e.g. hard hat, safety harness, hearing protection, etc.	Major

Figure 5.7 Hierarchy of risk control.

(Adapted from Safe Design Australia 2013, p. 19.)

Review of Case 2: detailed design of a commercial building

The second case-study building comprises seven storeys with a plant room located on the rooftop and a footprint area of 22 m x 20 m. The building houses mostly office spaces, with retail spaces and a foyer on the ground floor. The layout of all office floors is identical except for the sixth floor, which consists of office spaces and a terrace overlooking a nearby park. Figure 5.8 shows a perspective view of the complete building, while Figure 5.9 illustrates the ground floor plan.

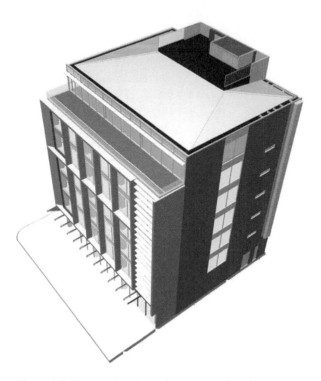

Figure 5.8 Perspective view of case study building 2.

(Image courtesy of Ganellen.)

The primary building structure is constructed of prefabricated structural steel components (stanchions and I-beams). Secondary structural elements are constructed of both precast and in-situ reinforced concrete (e.g. composite floors, staircases and lift shafts), and structural steel components (e.g. roof). The pitched roof is constructed of structural steel with a coloured steel-sheet covering. The facades comprise primarily glazing and non-structural precast concrete panels, which are supported by the structural slab at floor levels. Internal partitions are formed of Tasmanian oak. The floors of the building are finished with granite tiles, while the suspended ceiling is made of standard GIB board.

The building design had already incorporated several safe design options, such as significant use of prefabricated components, and the terrace on the sixth floor has balustrades of 1.10 m high. The participant group that

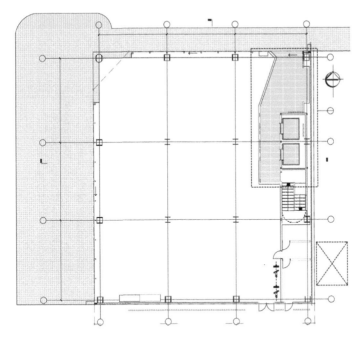

Figure 5.9 Ground-floor plan of case study building 2.

(Image courtesy of Ganellen.)

analysed the designs discovered ten critical design issues that are conducive to fall risks, within the limited time available. Table 5.2 shows the PtD review results.

Reflections

The initial survey that assessed the participants' proficiencies in PtD demonstrated that they were not highly skilled. However, with the support of the mobile app, they produced design review reports of high standard, as expounded in the preceding section. Moreover, at the conclusion of the review exercise, the following question was repeated to the participants:

How would you rate your current level of skills/knowledge in making better design choices to prevent fall accidents during construction and maintenance of buildings (1 = low; 5 = high)?

| 1 | 2 | 3 | 4 | 5 |

Table 5.2 PtD review results for case study building 2

Design aspect analysed	Details of fall hazards identified	Risk assessment	Prevention through Design suggestion by the review panel	Type of risk control applied
Roof – currently a pitched roof with an elevation of 5° has been designed	Although the elevation of the roof is low, because the roof area is large, risks of slip and fall could occur The pitched roof is constructed and maintained at 28 m above the ground, which creates potential fall hazards Installing or maintaining rainwater gutters at 28 m above the ground level is very risky	Likelihood: *Possible* Impact: *Catastrophic* Risk rating: *Very high*	1. Substitute the pitched roof with a concrete flat roof that can be built as a composite floor like the other floors in the building; this will eliminate fall risks during both construction and maintenance of the roof and rainwater gutters 2. Design strong railings or parapet walls along the edges of the roof slab to a height greater than 1.10 m 3. Specify cast-in sockets along the roof slab edges to enable early installation of railings 4. Install permanent anchor points in the roof slab to provide strong tie-off points for lifelines during construction and maintenance	1. Substitution 2. Isolation 3. Engineering 4. Engineering
Rooftop mechanical plant	Installation and regular maintenance of the rooftop mechanical plant present serious fall hazards to workers during the lifetime of the building	Likelihood: *Unlikely* Impact: *Catastrophic* Risk rating: *Very High*	1. If practical, place the mechanical equipment on the ground, which will eliminate the need for work on the rooftop for installation and maintenance of the plant 2. If this is not possible, locate rooftop equipment well away from roof edges and floor openings	1. Elimination 2. Isolation

Item	Hazard	Risk assessment	Control measure	Control type
In-situ concrete staircases along the east side of the building on all levels	Construction of the in-situ concrete staircases will involve formwork, reinforcement fixing and concrete casting at different heights above the ground; these present numerous fall hazards	Likelihood: *Possible* Impact: *Major* Risk rating: *Very high*	Consider prefabricated staircases that can be erected as a single assembly on the ground, minimising work at heights; moreover, specify the balustrades to be erected as part of the prefabricated staircases. If prefabricated staircases are not practical, specify Stairmasters that come with pre-attached railings and cast concrete in them	Substitution
Lift shaft – reinforcement	Use of bar reinforcement for core walls poses numerous trip and fall hazards at heights	Likelihood: *Likely* Impact: *Minor to Moderate* Risk rating: *High*	Use prefabricated reinforcement fabrics for core walls. For safe joining of core wall reinforcement between different floor levels for continuity, allow vertical lapping reinforcement to extend 1.8 m above the finished floor level	Substitution
Lift shaft – openings	Wall openings and floor penetrations of the lift shaft and risers pose potential fall risks as they are present on all levels of the building	Likelihood: *Possible* Impact: *Catastrophic* Risk rating: *Very high*	Continue reinforcement bars through floor penetrations and openings in lift core walls, which can be cut and removed after work, eliminating falls through lift shaft openings	Isolation
Terrace on 6th floor –15 mm toughened glass balustrades of 1100 mm height fixed in to concrete with 90 mm rods and a 40 x 25 mm handrail is connected to the glass balustrades	Failure of terrace railing members during the occupancy can cause serious fall risks, likely leading to fatalities	Likelihood: *Possible* Impact: *Catastrophic* Risk rating: *Very high*	Design railings to support a minimum of 90 kg applied within 50 mm in any direction (downward/outward) and at any point along the railing. Design intermediate vertical members of balustrade with a maximum spacing of 475 mm	Substitution

(*continued*)

Table 5.2 (continued)

Design aspect analysed	Details of fall hazards identified	Risk assessment	Prevention through Design suggestion by the review panel	Type of risk control applied
Ceiling – 1200 x 600 mm suspended grid ceilings with 13 mm standard GIB board on 90 x 45 mm timber framing on all levels	Workers may fall down while installing or maintaining ceilings due to the failure of ceiling frames	Likelihood: *Possible* Impact: *Moderate* Risk rating: *High*	Specify ceiling hangers and connections (frames) that are capable of supporting construction live loads, including the weight of the worker	Substitution
External claddings – both glass and concrete wall panels on all sides of the building	Workers may fall off the edge of the building when installing or maintaining external claddings at heights	Likelihood: *Possible* Impact: *Catastrophic* Risk rating: *Very High*	Specify permanent anchor points in structural members along the perimeter of the building around claddings to provide firm connections to fall-arrest systems and scaffolds during installation or maintenance of external cladding	Engineering
Glass windows on upper floors	Fall accidents could happen during the installation and maintenance (including regular cleaning) of the windows	Likelihood: *Unlikely* Impact: *Catastrophic* Risk rating: *Very High*	1. Design windows that can be installed and maintained from inside the building 2. Make members of the windows and the glazing system strong enough to withstand accidental and casual loading equivalent to that which would occur if a person fell against them; this will help to reduce falls caused by failing window members when forces are applied on them by construction or maintenance workers	1. Substitution 2. Substitution
External doors on the south side of the building	No canopy or overhang is provided over these doors; this may cause slippery conditions due to rain, snow, dew, fog, etc. and thereby slip and fall risks for building occupants	Likelihood: *Possible* Impact: *Moderate* Risk rating: *High*	Provide a canopy of adequate size over the external doors on the south side of the building	Isolation

The two different responses given by the same participants are compared in Figure 5.10. The participants were also asked to comment on the following issues at the conclusion of the workshop:

- effectiveness of the functionalities in the mobile app to support building design/design review for fall accident prevention;
- benefits that the mobile app can offer to building designers;
- benefits of introducing a mobile app rather than a stand-alone system or web system;
- potential challenges for implementing the mobile app in the industry.

The comments offered are aggregated and summarised below.

Mobile app functionalities and performance

There was an undivided view amongst the workshop participants over the effectiveness of the app functionalities and its performance, which is reflected in the following direct quotes from them:

- I find the app really useful. It has been designed in a way that it can help me check out almost all possible fall hazards without missing any. The app is well organised to give me directions to check where I could have created a dangerous design and then fix it either by elimination, substitution, isolation or engineering control.
- The app invokes hazard/risk considerations in each stage of a building design process. Moreover, the functions of project-specific suggestions

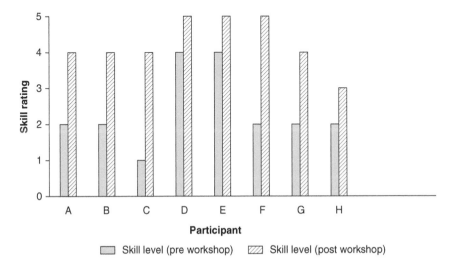

Figure 5.10 Comparison of skill levels.

and project questionnaire help the user to customise the use of the app according to their project characteristics.

- The PtD suggestions in the app are well-organised and easy to retrieve. Moreover, the images and illustrations are informative.
- The app's functionalities are reasonably effective for PtD review of designs and can help to ensure that all relevant hazards are considered. It also offers an easy and practical way to conduct design reviews. Moreover, no bugs were encountered during its use.
- The app facilitates the identification of unrecognisable safety hazards and indeed was very effective is guiding us to identify eight serious fall hazards in our conceptual design.
- The app was extremely effective in highlighting areas of design safety that could easily be missed. Having different building types and design options, and the risks associated with them, is also very useful. Moreover, the app is bug-free.
- By checking off the 28 questions in the project-specific suggestions section, which provides a report, consultants and contractors can create a checklist of items picked up by the app which were not addressed initially.

Benefits of the app to building designers

The participants indicated several potential business impacts of the app on building designers. The different benefits and potentials the app could offer, as commented, include the following:

- the app can help designers to exercise their duty of care and due diligence;
- it will ensure that clients will be provided with safe buildings throughout – safe during both construction and operation;
- it provides you with effective guidelines for reducing risks through design for different situations and types of building;
- checking my design against the suggestions in the app will ensure that I am producing as safe designs as possible;
- the app can help designers to be aware of the risk consequences of various design considerations and make their designs accordingly so that they eliminate or reduce risks;
- as safety in design has become a global concern, the knowledge offered by the app is important for designers to prevent hazards;
- the app would support designers in different design stages to mitigate risks during the construction process by providing ideas to improve their design to take account of safety;
- because PtD has become mandatory in Australia and in many other countries, the app will be a useful tool for design professionals to meet the legal requirements;

- having an app to support PtD enables designers to have access to up-to-date information/knowledge, provided that the app is kept updated regularly; in the case of using a textbook, the designer would have to take the initiative to find the relevant updates;
- it could help designers to understand safety hazards in construction better so that they could take them into consideration when designing.

Mobile app versus stand-alone/web system

All the participants unanimously held the view that the mobile app approach is better than a stand-alone or web system approach. Some key reasons for their view, as quoted by them, include the following:

- as an app, the system can run without the need for an internet connection and a registry of users; moreover, it is more cost-effective to buy an app for business purposes than registering to a web system or buying a stand-alone licence;
- it is handy and quicker to use your iPhone or iPad to check your design as it offers greater flexibility to check your design or hold design-review discussions regardless of your location;
- it gives flexibility in use as it is more mobile than a laptop or a computer unit;
- information will be readily available in your mobile;
- the app can be a quick learning tool as opposed to other modes.

Potential challenges

A couple of potential challenges in implementing the app in the industry were highlighted by the participants as below:

- the app needs to be updated regularly to maintain its effectiveness as new design features and hazards may surface in the future;
- some designers may not believe in the app and think that their systems are sufficient.

These findings collectively provide evidence as to the practical implications/business impacts of the new mobile app and the quality of information/knowledge offered by it. They also validate that the app-enabled mobile computing model for accident PtD is an effective approach.

App verification and fine-tuning

The use of the app during the workshop was smooth, and no bugs were encountered in the business logics and technical performance of the app. However, separate from the workshop, the app was checked meticulously by the researcher to identify any technical problems, and the following issues were detected.

a   The PtD suggestion pages looked too cramped. Spacing between para-graphs and bullet points was insufficient. Reformatting of the text in suggestions was needed for better readability, including justifying text, indenting bullet points, and allowing space breaks between bullets and paragraphs.
b   Images in suggestion pages could not be zoomed. Users should be able to use the standard 'two fingers' gesture to zoom in and out. This should apply to the whole page so that text would zoom as well.
c   Images in the PDF were blurry due to over-compression as well as being inconsistent in size. There was too big a space between text and images for some of the suggestions.
d   Tick and X buttons on questionnaire pages were not very sensitive and did not always respond to a tap. X was particularly unresponsive.

The mobile app was then fine-tuned to address these technical issues.

Conclusion

The evaluation exercise aimed to examine both the conceptual and techni-cal soundness of the mobile app for supporting fall PtD. The app was evaluated against four sets of criteria via a case-study-driven charrette workshop. The criteria comprised: (1) business impacts (benefits and oppor-tunities), (2) functionalities (accessibility, modularity and user friendliness), (3) information quality (completeness and usefulness), and (4) performance (speed, responsiveness and stability). The evaluation demonstrated that the app can offer significant business benefits to building designers by helping them to perform their duty of care and due diligence under the Workplace Health and Safety legislation. The functionalities, performance and infor-mation offered by the app were also found appropriate and sound to support a building designer's role in accident PtD. Moreover, the mobile computing app approach to decision support was favoured over a stand-alone or a web system. In conclusion, it was established that the app serves the intended purpose effectively.

References

AS/NZS 4360. (1999) *Risk Management*. Homebush, Australia: Standards Australia.
Dařena F. (2011) Information systems evaluation criteria based on attitudes of grad-uate students. *Journal of Efficiency and Responsibility in Education and Science*, 4(1): 46–56.
Freeman R E. (1984) *Strategic Management: A Stakeholder Approach*. Boston: Pitman Publishing.

Heo J and Han I. (2003) Performance measure of information systems (IS) in evolving computing environments: An empirical investigation. *Information & Management*, 40(4): 243–256.

Jamo Solutions. (2013) *Performance Testing and Monitoring of Mobile Applications*. URL (accessed 6 Jan. 2014): http://www.jamosolutions.com/wp-content/uploads/2013/08/White-Paper-M-eux-Test-Performance-Testing-of-Mobile-Applications.pdf.

Kamardeen I. (2009) *Controlling Accidents and Insurers' Risks in Construction: A Fuzzy Knowledge-based Approach*. Nova Science: New York.

Safe Design Australia. (2013) *Safety in Design Manual*. Unpublished consultancy report prepared for Tzannes Associates. Tzannes Associates: Sydney.

Serafeimidis V and Smithson S. (2000) Information systems evaluation in practice: A case study of organisational change. *Journal of Information Technology*, 15(2): 93–105.

Smith N D. (2012) *Design Charrette: A Vehicle for Consultation or Collaboration*. URL (accessed 12 Feb. 2014): http://www.kwokka.com.au/assets/nicola-d-smith-design-charrette.pdf.

Sojda R S. (2007) Empirical evaluation of decision support systems: Needs, definitions, potential methods and an example pertaining to waterfowl management. *Environmental Modelling & Software*, 22(2007): 269–277.

Stockdale R and Standing C. (2006) An interpretive approach to evaluating information systems: A content, context, process framework. *European Journal of Operational Research*, 173(2006): 1090–1102.

Yusof M M, Paul R J and Stergioulas L. (2005) Health information systems evaluation: A focus on clinical decision support system. In: Engelbrecht R, Geissbuhler A, Lovis C and Mihala G (eds), *Connecting Medical Informatics and Bio-Informatics: Proceedings of MIE2005: The XIXth International Congress of the European Federation for Medical Informatics*, Amsterdam: ISO Press, pp. 855–860.

6 Accident Prevention through Design curriculum design for higher education

Introduction

It has been established in previous studies that promoting research and education is one of the essential strategies for facilitating accident Prevention through Design (PtD) adoption in the construction industry. Two action items were suggested for this strategy, namely: expanding the existing body of knowledge around accident PtD through research; and integrating accident PtD course materials into design and engineering curricula in tertiary education institutions. Preceding chapters in the book implemented the first action item in the context of fall prevention. This chapter, utilising the newly built knowledge base, demonstrates how PtD can be incorporated into design and engineering curricula in tertiary institutions. The chapter first discusses the importance of facilitating deep learning via constructivist pedagogy and then showcases a sample of constructivist curriculum design for fall PtD. The chapter continues by explaining how assessments could be utilised to enhance student learning and then offers model assessment tasks for such purposes. Finally, some online resources that can assist PtD curriculum development are outlined.

Teaching for deep learning

Fry, Ketteridge and Marshall (2009) defined learning as a process of change in that learners create knowledge and mental models of their understanding of the world, which can help them develop behaviours appropriate to specific situations. The process involves acquiring factual information, skills, techniques and approaches that can be retained and used as is appropriate; making sense or abstracting meaning; interpreting and understanding reality (Smith 2003). Eklund (1995) identified two broader paradigms of learning, namely:

- behaviourist learning paradigm – considers learning as an organised transfer of knowledge with a structured learning strategy and the learning is teacher-centred;

- constructivist learning paradigm – views learning as the formation of mental models or constructs of understanding by the learner, supported by experiential learning and group work and the learning is student-centred.

The behaviourist paradigm can be regarded as the conventional approach that has been used for centuries, while the constructivist paradigm is a modern approach that has gained popularity in higher education.

Depending on whether teaching is teacher-centred or learner-centred, students may adopt either a surface approach or a deep approach to learning. The surface approach to learning, also referred to as rote learning, promotes task completion through memorisation and recall of information. The deep approach, on the other hand, leads learners to understand and seek meaning of concepts in context by critical evaluation and relating new ideas to existing understanding. It has been proven by educational research that through deep learning students achieve the required learning outcomes effectively. The learning outcomes, also known as graduate attributes, include: creativity, problem-solving skills, professional skills, communication skills, teamwork, and lifelong learning, which should be contextualised in programmes and courses that students undertake (Biggs and Tang 2011). Prosser and Trigwell (1999) claimed that the student-centred, constructivist approach to teaching will encourage deep learning.

Because the student approach to learning and the development of graduate attributes are heavily influenced by the teaching practice of lecturers, they must make sure that curriculum design, choice of teaching and learning methods, and assessment tasks support deep learning of concepts by students. Fry, Ketteridge and Marshall (2009) argued that learning requires opportunities for practice and exploration, space for thinking and critical reflection, interaction and collaboration with peers, and learning from peers and experts. In order to achieve these in practice, Kember and McNaught (2007) suggested a variety of active learning strategies, in addition to the most common modes of university teaching such as lectures, seminars and tutorials, which include: (1) mini research projects, (2) case-based teaching, (3) problem-based learning, (4) role play, (5) reflective journals, (6) experiential learning, (7) peer tutoring, (8) games and simulations and (9) computer-enhanced learning.

Constructivist curriculum design for fall PtD

The preceding section established that a constructivist curriculum is conceivably the effective path for teaching because this can help nurture students with practical skills essential for fall PtD in construction. This section proposes exemplary learning contents and activities for fall PtD education in design and engineering degree programmes, which can facilitate deep learning. PtD might be incorporated in one of two ways into existing design and engineering curricula.

1   PtD adoption:
    Adopt PtD materials and learning activities into existing modules
    around workplace health and safety, risk management, etc. and enhance
    subject design.
2   PtD adaptation:
    Integrate safety or PtD as one of the key principles such as those of
    functionality, cost, sustainability, aestheticism, etc. that need to be met
    in the design and engineering of facilities, and modify marking criteria
    for assessable tasks by encompassing criteria relating to safe design.

In either model, an interdisciplinary collaboration between educators is
likely to produce better learning and teaching outcomes.

Sample module design for fall PtD

Regardless of the curriculum model applied, the content for PtD education
should present threshold concepts prior to discussions on how designs can
be reviewed and revised to take account of PtD principles. Table 6.1 shows
the threshold concepts of PtD education in the upper part of the table and
the focused topics for fall prevention in the lower part of the table.
Resources relating to the contents of the table can be found in the preceding
chapters of the book and educators may refer to the appropriate chapters
for preparing elaborative teaching materials. In addition, the mobile app
can be used as a powerful learning and teaching support tool.

Table 6.1 Model subject design for fall PtD

Threshold concepts of PtD education:

- safety profile of construction and the contribution of design to accidents;
- regulatory frameworks and the duty of care of designers for safety;
- PtD process and tools (PtD review framework, risk-assessment matrix, risk-control hierarchy and hazard checklists and databases);
- business benefits of PtD.

Focused topics (e.g. fall PtD in building construction):

- fall prevention at the early design stage;
- fall prevention at the detailed design stage;
  - Safe foundation design
  - Safe building frame design
  - Safe staircase and railing design
  - Safe ramp and passage design
  - Safe door and window design
  - Safe walling and finishing
  - Safe roof design
  - Safe skyrise greenery design
  - Safe steel structure design.

The course delivery strategy for the contents may be planned as a combination of lectures/seminars, case analyses, literature reviews, portfolios, and group discussions and critical thinking. Lecture/seminar sessions may be the medium for the delivery of basic contents while tutorial sessions or student-directed learning could be employed for other activities.

Sample learning activities

Listed below are sample learning activities that lecturers might utilise for tutorial sessions or for initiating student-directed learning.

1 Identify Acts and Regulations in your jurisdiction that are related to workplace health and safety in construction, and explore the extent of duty of care of designers with regard to safety as assigned by those Acts and Regulations.
2 Compile a portfolio of Codes of Practice, Standards and Best Practice Guides for Safe Design of Buildings.
3 Compile a comprehensive checklist of keywords/terminologies with their definitions/explanations, which can guide designers in effectively identifying hazards when reviewing building designs for PtD purposes.
4 Locate a past, real-world professional-negligence lawsuit case that investigated the designer for a breach of duty of care for safety, and then analyse:

   a   the nature of failure to exercise due diligence and the breach of duty of care involved;
   b   the legal consequences of the professional negligence for the concerned designer;
   c   the risk-management measures that were available to the designer for protection against civil liabilities under the law of tort.

5 Your university is starting a new building project. Discuss how the effective adoption of PtD in the project would improve the business value of:

   a   the university;
   b   the builder;
   c   the designer.

6 Conduct a literature review of materials such as textbook chapters, journal and conference articles, and internet resources to explore:

   a   barriers facing effective adoption of PtD practices in the construction industry;
   b   practical strategies that may be utilised to overcome the challenges and expedite PtD adoption in industry;
   c   potential roles and contributions of different stakeholders such as clients, governments, professional institutions, tertiary institutions and industry bodies in operationalising the aforementioned strategies.

Enhancing student learning with assessment

Kamardeen (2014) postulated that assessment plays a large role in student learning and therefore can be utilised as a powerful driver of learning if a constructive alignment of learning outcomes, course contents and assessment tasks is maintained in curriculum design. Some other researchers also held a similar view. Elton and Johnston (2002) and Lombardi (2008) found in their research that students devote less than 10% of their study time to non-assessed learning tasks because they consider assessments as more important. Likewise, Kember and McNaught (2007) and Brown (2004) argued that students are motivated largely by assessment, and that assessment is possibly the best tool that lecturers could utilise to enhance student learning.

Higher-education literature categorises assessments into two types: summative (traditional) assessments, and formative (alternative) assessments. The summative approach aims to judge a student's competency/proficiency at the completion of an instructional unit or part of it. The most common summative assessment tools are a mid-unit quiz/test; an end-of-unit quiz/test/exam; a final essay/paper/drawing; a final presentation; a final project; etc. Formative assessments instead inform students of their learning progress towards some expected learning outcomes, and offer help and guidance to strengthen their areas of weakness and thereby improve their performance. Examples of formative assessments include problem-based learning; case studies; scenario-based learning; literature reviews; investigative project-based learning; portfolios; learning logs/journals; simulations/role plays; etc.

Law and Eckes (1995) claimed that summative assessments are an easy tactic for evaluating student performances as they are more objective and consistent. However, several disadvantages of summative assessments have been reported in the literature. Bailey (1998) critiqued traditional summative assessments as inauthentic and decontextualised, homogeneous, single-occasion tests, speed-centred and in which no meaningful feedback is provided to learners. Simonson et al. (2000) stated that summative assessments evaluate only learners' low-order cognition and thinking skills such as memorisation and recall. Law and Eckes (1995) added that this form of assessment can measure only learners' ability to perform a task at a given time and that the test score is not a good indicator of student progression.

Formative assessments, in contrast, evaluate higher-order cognition and thinking abilities as well as problem-solving skills (Reeves 2000). In this form of assessment, students have adequate opportunities to demonstrate their ability; therefore, this tactic is more student-growth focused. For example, if a student fails to perform in a task at a given time, they would still have the opportunity in a different scenario and time (Dikli 2003). Simonson et al. (2000) discussed many other benefits of formative assessments.

Because formative assessments are designed by simulating real-world scenarios, they provide students with opportunities to practise authentic tasks, apply their knowledge and skills to real-life settings, and work collaboratively. Moreover, lecturers are able to gain better insights of student learning. Davies (2000) posited that formative assessments offer deep-learning platforms as they promote active student engagement, as well as encouraging them to take risks, and learn by doing and making mistakes. However, formative assessments demand more time and effort from both the lecturer and students (Law and Eckes 1995).

Kamardeen (2014) suggested the use of an integrated assessment scheme in that: (1) formative tasks could be utilised for team-based learning in real-world contexts and for leveraging learning by continual feedback; and (2) summative assessments such as online quizzes could be introduced after lectures to test students' grasp of lecture contents, so that students' interest in the lecture could be retained and thereby the understanding of the topic could be improved. This would enable the lecturer and students to gain the benefits of both approaches.

Model assessment tasks for fall PtD

This section presents model assessment tasks that may be utilised by PtD educators to form an integrated assessment scheme in their modules. The samples include both summative and formative tasks, which can be mixed and matched for a given module.

Model Task 1: PtD audits of buildings

Form groups of four, conduct a PtD audit of an existing building within your campus premises and report your findings. No two groups can study the same building. Each group is required to perform the following tasks:

- carefully study the different locations and elements in the chosen building and identify fall hazards posed by their designs to building users and maintenance workers;
- take pictures of the hazardous designs and annotate them with explanations for effective reporting purposes;
- provide alternative design suggestions for the identified elements and locations; it is suggested that students use diagrams and sketches to clearly communicate their safe design proposals;
- make a presentation to the class of your findings and proposals on a date specified by the lecturer, receive feedback from your peers and the lecturer, and improve your work to take account of the feedback;
- submit a final report of your study to the lecturer in accordance with the assessment-submission protocols.

NB: It is critical that groups obtain any approvals required from the building manager before undertaking the study. If ethics approval is required from the university to carry out such a study, students must obtain such approvals prior to commencing the task.

Model Task 2: PtD review of a preliminary design for a house

Figure 6.1 illustrates the preliminary design that has been received by a client from a home builder for her dream house. The house is to be built on a block of flat land and the proposed design includes the following features:

- number of storeys – 3;
- separate living and dining areas on the ground floor;
- master bedroom with en suite;
- number of additional bedrooms – 2;
- study room – 1;
- single garage;
- roof terrace for greenery and relaxation.

Moreover, the builder proposes to construct the house with an in-situ concrete frame (i.e. pad footings, columns, beams, slabs and staircases) and brick masonry walls for the enclosure (200 mm thick) and internal dividers (100 mm thick). Brick masonry walls on the ground floor will be supported by strip footings. The floor on all levels is to be finished with ceramic tiles, except for the garage and rooftop greenery areas, which will be rendered, and plastering and painting will be the finishing materials for the walls and soffits of slabs.

Students are to form groups of four, play the role of the design review team and prepare a preliminary-design fall PtD review report for the project. The report will include the following:

- identification of hazards in the proposed design that present fall risks for building workers, occupants of the house, or maintenance workers, with clear annotations and explanations;
- alternative design suggestions for the fall hazard-prone locations and elements in the proposed design so that the hazards are eliminated or controlled effectively; it is strongly advised that groups use suitable sketches and diagrams with annotations to communicate their design suggestions effectively;
- submission of your PtD review within your report to the lecturer in accordance with the assessment-submission protocols.

Model Task 3: scenario analysis

For each of the following design scenarios, identify the fall hazards inherent in them that have the potential to cause risks to construction workers,

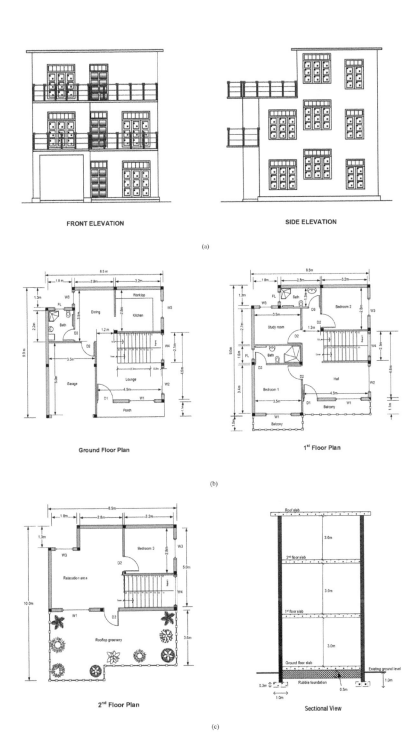

FRONT ELEVATION   SIDE ELEVATION

(a)

Ground Floor Plan   1st Floor Plan

(b)

2nd Floor Plan   Sectional View

(c)

Figure 6.1 House design for fall PtD review.

Figure 6.2 Air-conditioning system for a house.

maintenance workers and/or users, and suggest alternative design options to eliminate or reduce the risk(s). Use sketches, if applicable, to clearly illustrate your ideas/suggestions.

1   Scenario 1: Air-conditioning system for a two-storey free-standing house (see Figure 6.2).
2   Scenario 2: Ramp in a local council building with handrails of 900 mm high from the ramp surface. The vertical posts supporting the handrails are spaced at 1200 mm intervals (see Figure 6.3).

Figure 6.3 Ramp in a local council building.

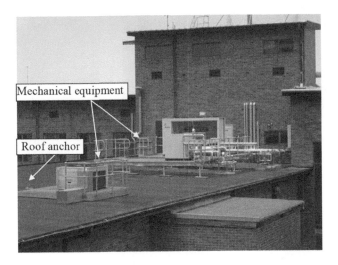

Figure 6.4 Rooftop equipment.

3   Scenario 3: Permanently placed mechanical equipment on the rooftop of a building. The rooftop has permanently fixed roof anchors along the middle line (see Figure 6.4).

4   Scenario 4: Escalator and a staircase within an institutional building are illuminated by permanently fixed lights above them. The staircase's width between faces measures 1.5 m, while the escalator's width is 0.75 m. The escalator is located right next to the staircase and lights are fitted above (see Figure 6.5).

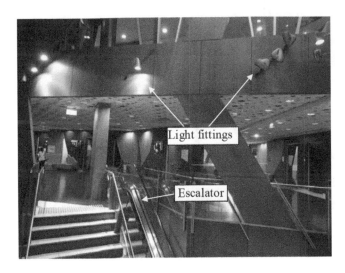

Figure 6.5 Light fittings for stairs.

Figure 6.6 Glass roof panels.

5   Scenario 5: Glass roof panels for the roof of a passage to provide natural lighting to the passage area (see Figure 6.6).
6   Scenario 6: Projected building on a university premises (see Figure 6.7).

Figure 6.7 Projected building.

Model Task 4: PtD quiz

This section contains summative questions that assess students' learning of PtD concepts, which may be utilised to complement lectures in the form of end-of-lecture quizzes or for end-of-session test purposes. The questions were adapted from ASCC (2006, Part 2C pp1-13).

Question 1 – Matching

Pair up the terms below with suitable definitions.

Terms:

Hazard	Risk	Accident
Prevention through Design	Risk matrix	Duty of care
Non-compliance	Professional negligence	Hierarchy of control

Definitions:

- A state or set of conditions that has the potential ....
  to lead to an accident
- An undesired and unplanned event that results ....
  in a loss
- A variety of risk-control options that are used to ....
  manage workplace health and safety risk
- Doing everything reasonably practicable to protect ....
  the health and safety of workers
- Failure to exercise due diligence ....

Question 2 – True/False

You have designed and overseen the construction of a playground for the local school – services that you carried out free of charge. Since you are not being paid for your professional services, you are absolved from any duty of care.

☐ True
☐ False

Question 3 – Multiple answers

When an injury occurs on a building site, who can be subjected to civil or criminal action?

☐ The building owner
☐ The designer
☐ The builder
☐ Workers

Question 4 – Organising

Arrange the below-listed elements of the workplace health and safety regulatory framework in descending order of influence on design and engineering practice (1 = most influential; 5 = least influential).

☐ Best-practice manuals
☐ Standards
☐ Codes of practice
☐ Regulations
☐ Acts

Question 5 – Multiple answers

Compliance with which of these elements of the workplace health and safety regulatory framework is not mandatory?

☐ Regulations
☐ Standards
☐ Codes of practice
☐ Acts
☐ Best-practice manuals

Question 6 – Multiple answers

A designer who has been found to have not complied with the safety duty of care could face which of the following:

☐ Criminal legal action
☐ Civil legal action
☐ Disciplinary action by the employer
☐ Disciplinary action by the relevant professional association

Question 7 – True/False

Controlling safety risk in construction projects through the use of personal protective equipment is always effective.

☐ True
☐ False

Question 8 – Multiple choice

A designer is proposing a steel post for a school playground, instead of timber, to reduce the risk of splintering. What is the type of hazard control applied in this case?

- ☐ Elimination
- ☐ Substitution
- ☐ Isolation
- ☐ Engineering

Question 9 – True/False

Controlling safety risk in construction projects through the use of administrative controls (e.g. providing training and implementing strict site-safety protocols) is always effective.

- ☐ True
- ☐ False

Question 10 – True/False

Controlling safety risk in construction projects through the use of both administrative controls (e.g. providing training and implementing strict site-safety protocols) and engineering technology is always effective.

- ☐ True
- ☐ False

Question 11 – Multiple choice

When is adherence to a certain work practice a legal requirement?

- ☐ When it is referred to in a Code of Practice
- ☐ When it is referred to in an Act or Regulation
- ☐ When it is prescribed in a Standard
- ☐ All of the above

Question 12 – Organising

The hierarchy of risk control is a variety of risk-control options that are used to manage workplace health and safety risks effectively. Order the below-listed options, with 1 for the most protective and therefore most preferred option, and 5 for the least protective and therefore least preferred option.

- ☐ Administrative (procedural) controls
- ☐ Engineering controls
- ☐ Elimination
- ☐ Personal protective equipment
- ☐ Substitution

Additional online resources

Many useful online resources are available on the internet, which can assist design and engineering educators in developing effective course contents, learning activities and assessments. A few such resources are listed below:

- *Design for construction safety:*
  This website may be considered as one of the primary reference sites for PtD. The site has materials covering PtD principles, best practice and past design-fault cases as well as toolkits and references/links to many other resources.
  URL: http://www.designforconstructionsafety.org/index.shtml

- *Design best practice:*
  This website includes case studies, best-design practices, bad-design practices and other PtD information and resources.
  URL: http://www.dbp.org.uk/welcome.htm

- *National Institute for Occupational Safety and Health (NIOSH):*
  The PtD sub-domain of the website of NIOSH contains several PtD-related materials that can be utilised for education purposes.
  URL: http://www.cdc.gov/niosh/topics/ptd/default.html

- *Engineering failures: case studies in engineering:*
  This website serves as a source of case studies in engineering failures whereby readers can reflect upon the root causes and obtain insights and learn lessons.
  URL: http://engineeringfailures.org

Conclusion

Higher education is regarded as an influential means that is capable of ingraining PtD principles as one of the key design and engineering criteria and thereby promoting its wider adoption in industry. Given that PtD is a relatively new concept, model course designs and assessment tasks should be made available to enable the integration of PtD into existing design and engineering curricula. This chapter has helped to achieve this objective and it is believed that the contents of this chapter could have positive implications for the tertiary education sector and subsequently for industry and society as a whole.

References

Bailey K M. (1998) *Learning About Language Assessment: Dilemmas, Decisions, and Directions.* Boston: Heinle & Heinle.

Biggs J and Tang C. (2011). *Teaching for Quality Learning at University* (4th edn). Berkshire: Open University Press.

Brown S. (2004) Assessment for learning. *Learning and Teaching in Higher Education, 2004–05*(1): 81–89.

Davies A. (2000) *Making Classroom Assessment Work*. Merville, BC: Connections Publishing.

Dikli S. (2003) Assessment at a distance: Traditional vs. alternative assessments. *The Turkish Online Journal of Educational Technology*, 2(3): 13–19.

Eklund J. (1995) Adaptive learning environments: The future for tutorial software. *Australian Educational Computing*, 10(1): 10–14.

Elton L and Johnston B. (2002) *Assessment in Universities*. URL (accessed 5 Oct. 2013): http://eprints.soton.ac.uk/59244/1/59244.pdf.

Fry H, Ketteridge S and Marshall S. (2009) *A Handbook for Teaching and Learning in Higher Education* (3rd edn). Abingdon: Routledge.

Kamardeen I. (2014) Stimulating learning with integrated assessments in construction education. *Australasian Journal of Construction Economics and Building*, 14(3): 86–98.

Kember D and McNaught C. (2007) *Enhancing University Teaching: Lessons from Research into Award-Winning Teachers*. Abingdon: Routledge.

Law B and Eckes M. (1995) *Assessment and ESL*. Manitoba, Canada: Peguis Publishing.

Lombardi M M. (2008) *Making the Grade: The Role of Assessment in Authentic Learning*. URL (accessed 8 Oct. 2013): http://net.educause.edu/ir/library/pdf/eli3019.pdf.

Prosser M and Trigwell K. (1999) *Understanding Learning and Teaching: The Experience in Higher Education*. Buckingham: The Society for Research into Higher Education and Open University Press.

Reeves T C. (2000) Alternative assessment approaches for online learning environments in higher education. *Educational Computing Research*, 3(1): 101–111.

Simonson M, Smaldino S, Albright M and Zvacek S. (2000) Assessment for distance education. *Teaching and Learning at a Distance: Foundations of Distance Education*. Upper Saddle River, NJ: Prentice-Hall.

Smith C. (2003) Experiential learning, diversity, and shared control: Doing civic education in metropolitan Detroit. *International Civic Education Research Conference*, 16–18 November 2003.

7 Conclusion

Introduction

This chapter concludes the book by highlighting the key findings of the research study and their implications. First, it provides a concise view of the research problem addressed. Then, the key findings and their contributions are elaborated. Following that, the educational, practical and research-methodological implications of the contributions are explained. Finally, the chapter lays the foundation stones for future research directions in the domain of accident PtD in construction.

Research summary

Falls are a leading cause of workplace fatalities, and serious injuries that result in permanent disabilities, among construction workers globally. Accident statistics provide evidence that falls are responsible for about 30–50% of fatalities on construction sites and cost the construction industry millions of dollars every year, creating a distressing socio-economic burden globally. Prevention of fall risks is therefore a key priority in construction.

Building design concepts and choices made by architects and engineers are criticised as a key causal factor for these falls in about 40–50% of circumstances. Conversely, better designs can significantly reduce falls in construction. Prevention through Design (PtD) is therefore recognised by the construction industry and researchers as an effective means to improve site safety as it helps to eliminate hazards at source. Realising its potential to curtail the number of falls and other construction accidents in general, many governments around the globe have mandated PtD in recent times. This legislative change has redefined the roles and responsibilities of designers in that they are now required by law to exercise PtD in their design practices. In other words, designers such as architects and engineers may be held liable for accidents that happen in projects designed by them. Whilst this change is healthy for the construction industry and its operatives, it gives rise to concern in designers because of the gaps in knowledge and skills they are encountering. Having primarily trained in design principles and not having been exposed to construction processes, designers have limited understanding of the hazards and risks triggered by their design concepts and choices. This impedes the successful adoption of PtD for fall prevention. It is therefore essential that designers have access to such integrated knowledge and decision-support tools if curtailing the number of lethal fall accidents in construction is to be achieved.

To this end, this study aimed to harness the power of the mobile computing technologies to provide designers with modular and just-in-time PtD decision support. It was decided to develop a mobile app as the PtD decision-support tool because mobile apps in general are: simple and easy to use by designers with any level of ICT competency; cost-effective; and offer wider accessibility compared to web systems or stand-alone decision-support systems. Moreover, this technology is less likely to face resistance within an organisational context due to its desirable qualities.

The research demonstrated the utility of the idea in the context of building designs, anchored on well-defined research methods. The various research techniques used at different stages of the study include: content analyses, rapid app prototyping and industry reviews, final app modelling and programming, and case-study-driven charrette workshops and technical testing for app evaluation.

Research findings and contributions

The research, through its findings, enhances the existing body of knowledge, as well as adding a new technological perspective to accident PtD. The study claims two primary findings, namely:

- a knowledge base of safe design solutions for fall prevention and control in building projects, which is detailed in Chapter 3;
- a working mobile app for fall PtD and its conceptual and logical designs, as expounded in Chapters 4 and 5. The developed mobile app can be found in Apple Store with the search keyword 'Fall PtD' and be installed on Apple devices such as iPhones and iPads for free.

The first finding addresses a significant gap in the body of knowledge for accident PtD in construction and satisfies the prerequisite for achieving the ultimate aim set out for the research. Despite PtD being recognised as an effective way of reducing the number of accidents on construction sites, and the fact that many governments and organisations around the world have been promoting it vigorously, its adoption in industry hitherto has faced challenges. Primary challenges are the still-evolving body of PtD knowledge and the PtD skills of designers. These two challenges are interrelated in that a developed body of PtD knowledge is likely to enrich the PtD competencies of designers. By establishing a new knowledge base of safe design solutions, the research contributes to alleviating the challenges facing building designers with regard to PtD knowledge and skills.

The second finding of the research takes PtD to the arena of mobile computing for the first time ever, thereby achieving the ultimate aim of the research. The finding proves to be a solid software toolkit that supports fall PtD in building projects. It is also an exemplary case and leads the way in demonstrating how conceptual and logical models can be developed and translated into cost-effective mobile apps to provide modularised, just-in-time decision support with increased flexibility in time and location. By building the mobile app solution, the research is facilitating faster adoption of PtD at a wider level as the app can be easily distributed to the international community via Apple Store.

Small-sized design-and-construction firms account for a significant proportion of the overall industry. It is therefore vitally important that PtD is embraced by these firms too if a successful adoption is to be achieved. Nonetheless, their interest in PtD-related investments seems marginal because of their scale, slim profit margins and lack of competent staff. Hence, this section of the industry should be supported by providing simple, straightforward and cost-effective solutions for PtD. The mobile app developed in this research is cost-effective and an easy technology, which can be operated by designers with any level of ICT competency. Moreover, it does not need expensive hardware or subscription/licensing and is highly flexible. These desirable qualities of the new innovation are able to take PtD to all levels of the industry quickly.

Practical implications

The mobile app could act as a catalyst for a faster adoption of PtD in design practices of all sizes, and help designers to exercise their duty of care and due diligence under the workplace health and safety legislation. Consequently, this can assist in producing safe-to-build designs in the industry, thereby reducing the number of construction accidents, particularly falls. This would discharge not only designers from their legal responsibility, but also clients and builders, because in the event of an accident, the current situation is that every one along the supply chain is likely to be held accountable and/or investigated. This causes productivity losses and other negative economic and business impacts for all the said parties.

Accidents cause numerous human sufferings and long-term social costs to victims and their families. These in turn become a significant socio-economic burden on national economies and society. Falls particularly create an excessive socio-economic drain, as has been established in the past workplace-accident and compensation statistics across many countries. By fostering a faster adoption of fall PtD and thereby reducing the incidence of falls, the mobile app could contribute to easing the socio-economic burden of nations as well as the sufferings of victims and their families.

Educational implications

The research findings yield a significant educational impact. Studies conducted by researchers around the world concerning practical strategies to overcome PtD implementation challenges suggest that leveraging research and education around PtD is a crucial step. This is further described as: expanding the existing body of knowledge around PtD through research; and developing design and engineering curricula in academic institutions, by integrating PtD course materials (Gambatese 2011; Manuele 2008; Schulte et al. 2008). In a similar vein, it is widely acknowledged by another set of researchers that PtD principles have not been effectively integrated into current design curricula (Zarges and Giles 2008; Cooke, Lingard and Blismas 2008; Zou, Yu and Sun 2009; Lingard, Cooke and Blismas 2012). This has not been possible due to the lack of a knowledge base and tools that integrate design, PtD and risk-management concepts. Nevertheless, the mobile-computing-enabled PtD methodology and the new knowledge base that emerges from this research could underpin the enhancement of design curricula at universities and other tertiary institutions.

Methodological insights for researchers

Apart from the primary findings and innovations, the research also contributes to the discourse of mobile app development methodologies. The supplementary contributions are as follows:

- it showcased, in detail, effective approaches for app planning, design and evaluation, which can be valuable for other researchers and be easily applied in their projects;
- the know-how for resolving some critical technical and procedural challenges that are likely to be encountered by app researchers during development endeavours is disseminated (refer to the section 'Lessons learnt' in Chapter 4).

The way forward

Studies undertaken by construction researchers around the world to identify prudent strategies to overcome PtD adoption challenges highlighted that research efforts were needed to advance the PtD body of knowledge with safe design methods and digital tools for design risk analysis and reviews. This book contributes to this international agenda, yet the focus of the book was limited to a specific part of the whole spectrum. This limitation, combined with the new direction set out by the book, opens up new needs and opportunities for further research. In this vein, the following future research is recommended in order to expand the PtD knowledge base in a similar fashion:

- The book is focused predominantly on fall PtD in building projects. However, fatal falls are suffered in all types of construction projects. It is therefore recommended to extend the study of fall PtD into civil engineering as well as into specialist trade services projects.
- Whilst falls are a primary cause of fatalities on construction sites, accidents attributed to strike-by-objects and electrocution are found to take a similar position as falls in terms of fatality consequences. New studies could be undertaken to create a PtD knowledge base for these risks and subsequently to develop mobile computing solutions.

PtD with Building Information Modelling (BIM) is another potential area for exploration. BIM is an emerging paradigm in the design and construction field, which enables designers to build a facility virtually on a computer, prior to building it physically on site. The resulting model is a data-rich, object-oriented, intelligent and parametric digital representation of a building that can be used to simulate construction and operation, and identify avenues for design optimisation (Rajendran and Clarke 2011). BIM is being used for many design and construction optimisations, including: design and visualisation, building code checking, indoor air-quality estimation, energy analysis, cost estimation, construction planning, and facilities management. However, the potential of BIM for PtD practices is largely unexplored. Future research is suggested to develop a BIM methodology/tool for fall PtD in construction. Such a tool could be developed to have the capability to audit a building model and then identify design aspects that are conducive to the possibility of falls. The PtD knowledge base created in this research can be utilised for the formulation of the rule base necessary for the proposed BIM tool.

References

Cooke T, Lingard H and Blismas N. (2008) ToolSHeDTM: The development and evaluation of a decision support tool for health and safety in construction design. *Engineering, Construction and Architectural Management*, 15(4): 336–351.

Gambatese J. (2011) *Findings from the Overall PtD in UK Study and Their Application to the US. Prevention through Design: A New Way of Doing Business: Report on the National Initiatives, Washington DC, 22–24 August, 2011.* URL (accessed 26 Apr. 2013): http://www.asse.org/professionalaffairs_new/PtD/Research%20Issues/John%20Gambatese.pdf.

Lingard H C, Cooke T and Blismas N. (2012) Designing for construction workers' occupational health and safety: A case study of socio-material complexity, *Construction Management and Economics*, 30(5): 367–382.

Manuele F. (2008) Prevention through Design (PtD): History and future. *Journal of Safety Research*, 39(2): 127–130.

Rajendran S and Clarke B. (2011) Building information modelling: Safety benefits and opportunities. *Professional Safety* (October): 44–51.

Schulte P A, Rinehart R, Okun A, Geraci C L and Heidel D S. (2008) National Prevention through Design (PtD) initiatives. *Journal of Safety Research*, 39(2): 115–121.

Zarges T and Giles B. (2008) Prevention through Design (PtD). *Journal of Safety Research*, 39(2): 123–126.

Zou P, Yu W and Sun A C S. (2009) An investigation of the viability of assessment of safety risks at design of building facilities in Australia. In: Ligard H, Cooke T and Turner M (eds), *Proceedings of the CIB W099 Conference 2009*, 21–23 October, Melbourne, Australia, CD-ROM, CIB W099, Paper No. 12.

Index